When
Elephants Weep

Also by Jeffrey Moussaieff Masson:

MY FATHER'S GURU:
A Journey Through Spirituality and Disillusion

FINAL ANALYSIS:
The Making and Unmaking of a Psychoanalyst

AGAINST THERAPY:
Emotional Tyranny and the Myth of Psychological Healing

A DARK SCIENCE:
Women, Sexuality and Psychiatry in the Nineteenth Century

THE ASSAULT ON TRUTH:
Freud's Suppression of the Seduction Theory

THE OCEANIC FEELING:
The Origins of Religious Sentiment in Ancient India

THE COMPLETE LETTERS OF SIGMUND FREUD
TO WILHELM FLIESS 1887–1904
(Editor)

THE PEACOCK'S EGG: Love Poems from Ancient India
(Translations by W. S. Merwin, Editor)

When Elephants Weep

The Emotional Lives of Animals

Jeffrey Moussaieff Masson

and Susan McCarthy

Delacorte Press

Published by
Delacorte Press
Bantam Doubleday Dell Publishing Group, Inc.
1540 Broadway
New York, New York 10036

Library of Congress Cataloging in Publication Data
Masson, J. Moussaieff (Jeffrey Moussaieff), 1941–
    When elephants weep : the emotional lives of animals / Jeffrey
  Moussaieff Masson, Susan McCarthy.
      p.   cm.
    Includes bibliographical references and index.
    ISBN 0-385-31425-6
    1. Emotions in animals.   I. McCarthy, Susan.   II. Title.
    QL785.27.M37   1995
    591.51—dc20   94-23819 CIP

Designed by Rhea Braunstein

Manufactured in the United States of America
Published simultaneously in Canada

June 1995

10   9

BVG

For the Fu
and Fiona

Contents

CONTENTS

Acknowledgments

In the process of researching this book we talked to many scientists, animal trainers, and others whose expertise was invaluable. We would particularly like to acknowledge the help of George Archibald, Mattie Sue Athan, Luis Baptista, Kim Bartlett, John Beckman, Mark Bekoff, Tim Benneke, Joseph Berger, Nedim Buyukmihci, Lisa De Nault, Ralph Dennard, Pat Derby, Ian Dunbar, Mary Lynn Fischer, Maria Fitzgerald, Lois Flynne, Roger Fouts, William Frey II, Jane Goodall, Wendy Gordon, Donald Griffin, David Gucwa, Nancy Hall, Ralph Helfer, Abbie Angharad Hughes, Gerald Jacobs, William Jankowiak, Marti Kheel, Adriaan Kortlandt, Charles Lindholm, Sarah McCarthy, David Mech, Mary Midgley, Myrna Milani, Jim Mullen, Kenneth Norris, Cindy Ott-Bales, Joel Parrott, Irene Pepperberg, Leonard Plotnicov, Karen Reina of Bristol-Myers Squibb, Diana Reiss, Lynn Rogers, Vivian Siegel, Barbara Smuts, Elizabeth Marshall Thomas, Ron Whitfield, and Gerald S. Wilkinson, among others, for their patience in talking to us. We are also grateful to Jennifer Conroy, Joanne Ritter, Mike Del Ross, and Kathy Finger of Guide Dogs for the Blind in San Rafael. Any errors we have made and any wild speculations, particularly those deemed scientifically disreputable, are not to be laid at their doors.

More personal thanks are owed to our friends and family also for their support and real assistance, especially Daniel Gunther, Joseph Gunther, Kitty Rose McCarthy, Martha Coyote, John McCarthy, Mary Susan Kuhn, Andrew Gunther, Barbara and Gerald Gunther, Thomas Goldstein, Martin Levin, and Bernard Taper; and Daidie Donnelley, Fred Goode, Justine Juson, Marianne Loring, Jane Matteson, Eileen Max, and Barbara Sonnenborn.

We also want to thank Elaine Markson for being a wonderful agent; Tony Colwell for having faith in this idea all along; Steve Ross for his enthusiasm and his indispensable help in making this the book it is; as for Kitty, only Kitty knows what Kitty is owed.

When
Elephants Weep

Prologue:
Searching the Heart
of the Other

"The Indian elephant is said sometimes to weep."
CHARLES DARWIN

Animals cry. At least, they vocalize pain or distress, and in many cases seem to call for help. Most people believe, therefore, that animals can be unhappy and also that they have such primal feelings as happiness, anger, and fear. The ordinary layperson readily believes that his dog, her cat, their parrot or horse, feels. They not only believe it but have constant evidence of it before their eyes. All of us have extraordinary stories of animals we know well. But there is a tremendous gap between the commonsense viewpoint and that of official science on this subject. By dint of rigorous training and great efforts of the mind, most modern scientists—especially those who study the behavior of animals—have succeeded in becoming almost blind to these matters.

I was led to my interest in animal emotions by experiences with animals—some traumatic, some deeply touching—as well as by the seeming opacity and inaccessibility of human feelings compared with their undiluted purity and clarity, at times, in my animal friends, and especially of animals in the wild.

In 1987 I visited a south Indian game reserve known for its wild elephants. Early one morning I set out with a friend to walk in the forest. After a mile or so we came across a herd of about ten elephants, including small calves, peacefully grazing. My friend stopped at a respectful distance, but I walked closer, halting about twenty feet away. One large elephant looked toward me and flapped his ears.

Knowing nothing about elephants, I had no idea this was a warning. Blissfully ignorant, as if I were in a zoo or in the presence of Babar or some other story-book elephant, I felt it was time to commune with the elephants. Remembering a Sanskrit verse for saluting Ganesha, the Hindu god who takes elephant form, I called *"Bhoh, gajendra"*—Greetings, Lord of the Elephants.

The elephant trumpeted; for a second I thought it was his return greeting. Then his sudden, surprisingly agile turn and thunderous charge in my direction made it all too clear that he did not participate in my elephant fantasies. I was aghast to see a two-ton animal come hurtling toward me. It was not cute and did not resemble Ganesha. I turned and ran wildly.

I knew I was in real danger and could feel the elephant gaining on me. (Elephants, I later learned in horror, can run faster than people, up to twenty-eight miles an hour.) Deciding I would be safest in a tree, I ran to an overhanging branch and leapt up. It was too high. I ran around the tree and raced into tall grass. Still trumpeting menacingly, the elephant came running around the tree in close pursuit. He clearly meant to see me dead, to knock me down with his trunk and trample me. I thought I had only a few seconds to live and was nearly delirious with fear. I remember thinking, "How could you have been so stupid as to approach a wild elephant?" I tripped and fell in the high grass.

The elephant stopped, having lost sight of me. He raised his trunk and sniffed the air, searching out my scent. Fortunately for me they have rather poor vision. I realized I had better not move. After a few long moments he turned away and raced off in another direction, looking for me. Soon I quietly picked myself up and, trembling, made my way slowly back to where my terrified friend had stood watching the whole episode, convinced she would witness my death.

Rudimentary knowledge of elephants would have kept me

safe: a herd with small calves is particularly alert to danger; elephants don't like their space invaded; flapping ears are a direct warning. The encounter itself was nothing but a projection of my own wish that a wild elephant would want to meet me.

It was wrong to think that I could communicate with a strange elephant under these circumstances. Yet he communicated very clearly to me: he was angry and I should leave. I believe this is a realistic description.

By contrast with animals, people's emotions are often distanced. For example, I experience heightened emotions in dreams —anger, love, jealousy, relief, curiosity, compassion—to a degree of intensity that is not paralleled in waking life. To whom do those emotions belong? Are they mine? Are they what I imagine a feeling to be like? In the dream there is nothing abstract about them: I *feel* extraordinary love, always for people for whom, in fact, I *do* feel love, just not to that degree. As a former psychoanalyst, I thought that these were feelings I had somehow repressed in my day life, and only had access to the real feelings in my night life. I theorized that the feelings were real, only access to them was barred. The feelings were always there, but could only become conscious at certain moments when some part of me was off guard—asleep, as it were. Somehow my ego had to be circumvented, an end run needed to be made, and they were there waiting, pure, unsullied, ready. Might animals have the more ready access to this feeling world that was largely denied my waking self?

Then there is the question of the feelings of others. What could be more interesting than what others *feel*? Do they feel the same things I do? I have found it hard to find out by talking, or even by reading. Songs, poems, literature, walking in the woods, evoke certain feelings. Sometimes they are strange, complex, inexplicable, even bizarre, often intense beyond comprehension. Where does this come from? I have long wondered. Why am I feeling this? *What* am I feeling? How could I name this?

In my training as a psychoanalyst, I discovered that analysts were not really all that interested in emotions. Or rather, they confined their interest to interpreting an emotion's meaning to the psyche or discussing whether an emotion was appropriate or inappropriate. I thought appropriateness was a ridiculous category. Emotions simply were. Moreover, they seemed to come unbidden. They were mysterious guests, hard to capture. Sometimes I thought I could feel something for only a brief second, or fraction of a second, but then it was gone and could not be recalled. Sometimes I would wake up in the middle of the night and remember a feeling I had once had and experience a kind of loss.

Psychoanalysis purports to be about feelings, especially deep feelings. For psychoanalysts the *essence* of a person is not what is thought or achieved, but what is *felt*. The standard, almost humorous, therapist question, "How do you feel about that?" turned out to be quintessential and hard to answer. We do not always know— hence the notion, early in Freud's work, of unconscious emotions, ones to which we are denied access. The early goal of psychoanalysis to make the unconscious conscious was directed toward bringing feelings into awareness, raising submerged emotions to the surface. Yet the question of emotions in dreams was and is barely touched in the psychological literature.

What fascinated me about animals was the ready access they seemed to have to their emotions. No animal, it seemed to me, needed to dream to feel. They demonstrated their feelings constantly. Annoy them, they have no hesitation in showing it. Please a cat, it purrs and rubs itself against you. What could appear as contented as a cat? A dog wags its tail and looks more genuinely pleased to see you than any human. What could appear as happy as a dog? Could anything seem as peaceful as a cow? Or are these merely human projections?

As a child, I had a duck that seemed to think I was its mother. It followed me everywhere. When we went on vacation, a neighbor offered to care for it. On our return, I eagerly asked how my duck was and he replied, "Delicious." I became a vegetarian that day. I still cannot bear to eat anything with eyes. The reproach is too deep.

I love dogs; it has always been clear to me that they lead extremely intense emotional lives. "No, Misha, no walk just now." *What?* The ears would cock. *Can I have heard right?* "Sorry, Misha, but no." Unmistakable. The ears flop. Misha would throw himself onto the floor. There was no mistaking the pure disappointment he was feeling. Just as unmistakable was his intense joy when I would say, "Okay, get your leash, we're going for a walk," and the sheer pleasure Misha felt on his walks, his delight at racing ahead, chasing leaves, doubling back, tearing off into the forest and returning behind and ahead of me. The contentment when we got home, built a fire, and I sat down to read, he to rest next to me, his face on my knee, was equally apparent. As he grew old, and could no longer walk as well, I could almost see him visit the scenes of his earlier life in his imagination. Nostalgia, in a dog? Well, why not? Darwin thought it possible.

In his book *The Expression of the Emotions in Man and Animals,* Charles Darwin had dared to imagine a dog's conscious life: "But can we feel sure that an old dog with an excellent memory and some power of imagination, as shewn by his dreams, never reflects on his past pleasures in the chase? and this would be a form of self-consciousness." Even more evocatively, he asked: "Who can say what cows feel, when they surround and stare intently on a dying or dead companion?" He was unafraid to speculate about areas that seemed to require further investigation.

Another reason I began thinking in some depth about animal emotions was the common experience of going to a zoo. We have all seen the look of forlorn sadness on the face of an orangutan, wolves pacing nervously up and down, gorillas sitting motionless, seemingly in despair, or perhaps having abandoned all hope of ever being free.

A pivotal book in my thinking about animal emotions was Donald Griffin's *The Question of Animal Awareness.* Attacked in many quarters upon its publication in 1976, it discussed the possible intellectual lives of animals and asked whether science was examining issues of their cognition and consciousness fairly. While Griffin did not explore emotion, he pointed to it as an area that needed investigation. Convincing and intellectually exciting, it made me want to read a comparable work on animal emotions, but I learned that there was almost no investigation of the emotional lives of animals in the modern scientific literature.

Why should this be so? One reason is that scientists, animal behaviorists, zoologists, and ethologists are fearful of being accused of anthropomorphism, a form of scientific blasphemy. Not only are the emotions of animals not a respectable field of study, the words associated with emotions are not supposed to be applied to them. Why is it controversial to discuss the inner lives of animals, their emotional capacities, their feelings of joy, disappointment, nostalgia, and sadness? Jane Goodall has recently written of her work with chimpanzees: "When, in the early 1960s, I brazenly used such words as 'childhood,' 'adolescence,' 'motivation,' 'excitement,' and 'mood,' I was much criticized. Even worse was my crime of suggesting that chimpanzees had 'personalities.' I was ascribing human characteristics to nonhuman animals and was thus guilty of that worst of ethological sins—anthropomorphism."

Hungry to learn more systematically about animal emotions, I found that the book I wanted to read was yet to be written. So I started researching accounts of specific animals.

Among the first people I asked about the emotional lives of animals were researchers working with dolphins. Dolphins show such delight in performing, even in creating new performances of their own, that an elaborate emotional component seems obvious. I visited Marine World Africa USA, near Berkeley, California, to meet Diana Reiss, a leading dolphin researcher. She showed me "her" four dolphins in their large, clean tank, all clearly eyeing her, watching her movements, eager for her to come into the water and play with them. I wanted to think they were happy, that they liked being there. I asked her. Oh, yes, she said, they feed, they mate, they are physically healthy, they enjoy the games she invents as part of her research. I nodded in agreement. But is that enough to count as happiness? I remembered what George Adamson, the husband of Joy Adamson of *Born Free* fame, said in his autobiography: "A lion is not a lion if it is only free to eat, to sleep and to copulate. It deserves to be free to hunt and to choose its own prey; to look for and find its own mate; to fight for and hold its own territory; and to die where it was born—in the wild. It should have the same rights as we have."

Thinking that experts who work with and study animals might offer observations in person that they would be reluctant to put into a scientific article, I asked other renowned scholars of dolphin behavior about their experience with the emotions their dolphins expressed. They were unwilling to speculate or even to offer observations. One said, "I don't know what *emotion* means." Another referred the matter to his female graduate students, implying that the subject was somehow beneath his scientific (or male?) dignity.

What these scholars said was undermined by what they did. One hugged his prize dolphin in a clearly emotional moment, at least for the researcher. The other could hardly leave at night, so attached had he become to what he called his "subjects." The female graduate students had many stories to tell about mutual affection between researchers and dolphins, even some free-living dolphins. It is hard to believe that these scientists would express intense feelings toward creatures they genuinely felt were emotionally insensate and could not return them or respond to them in any way.

In any event, how can anyone know that an animal feels nothing if the question has never been investigated? To conclude without study that it has no feelings or cannot feel is to proceed on a prejudice, an unscientific bias, in the name of science. This is not the only area in which scientists cling to unscientific dogma. Consider how long psychoanalysts denied the reality of child sexual abuse. Sexual abuse of children occurred long before Freud became interested in it, but his conclusion, without evidence, that it largely did not happen kept it hidden until the women's movement exposed its true prevalence.

Looking for information about how trainers worked with the emotions of the animals they used in their shows, I approached the public relations director at Sea World in San Diego. He told me bluntly that he disapproved of the notion of animal emotions and would not permit Sea World to be associated with my research because it "smacked of anthropomorphism." I was therefore astonished to see the shows there in which the killer whale and the dolphins were trained to wave, shake hands, and splash water at the spectators. They had been trained to behave like people—more precisely, like people who had been bent and formed into amusing slaves in the service of commercial exploitation.

Comparative psychology to this day discusses observable behavior and physical states of animals, and evolutionary explanations for their existence, but shies away from the mental states that are inextricably involved in that behavior. When such states are examined, the focus is on cognition, not emotion. The more recent discipline of ethology, the science of animal behavior, with its insistence on distinctions between species, also seeks functional and causal, rather than emotive, explanations for behavior. The causal explanations center on theories of "ultimate causation"—the animal pairs because this increases reproductive success—as distinguished from "proximate causation"—the animal pairs because it has fallen in love. Although the two explanations are not necessarily mutually exclusive—one of the best-known figures of ethology, Konrad Lorenz, spoke confidently of animals falling in love, becoming demoralized, or mourning—the field as a whole has continued to treat emotions as unworthy of scientific attention.

With the advent of laboratory studies of animals, especially in the 1960s, the distance maintained from the world of animal feelings became even greater. This distance supported scientists who do painful experiments on animals in believing that animals feel no pain or suffering, or at least that the pain animals feel is removed enough from the pain humans feel that one need not take it into account in devising experiments. The professional and financial interests in continuing animal experimentation help to explain at least some resistance to the notion that animals have a complex emotional life and are capable of experiencing not only pain but the higher emotions such as love, compassion, altruism, disappointment, and nostalgia. To acknowledge such a possibility implies certain moral obligations. If chimpanzees can experience loneliness and mental anguish, it is obviously wrong to use them for experiments in which they are isolated and anticipate daily pain. At the very least, this poses a matter for serious debate—a debate that has scarcely begun.

Some of the most innovative work being done with animals today is directed at language use, self-awareness, and other cognitive abilities, so that the willful blindness of science to the world of animal emotions seems on the verge of crumbling. The enticing subjects of cognition and consciousness are both easier to test and more respectable than that of emotion. Intelligence is certainly fascinating, but an animal, like a human, need not be intelligent to have feelings. What data there is on animal emotions comes not from laboratory work but from field studies. Some of the most esteemed animal researchers of our day, from Jane Goodall to Frans de Waal, from time to time defy orthodoxy and from their position of eminence within their fields use words like *love* and *suffering* to describe animals. Yet these aspects of their work are virtually ignored, and it remains professionally risky for less well established scientists to use such terms.

But there are signs of significant change. Recently Sue Savage-Rumbaugh, a scientist at the Yerkes Primate Center in Atlanta, Georgia, wrote in the preface to her book *Ape Language:*

> It is possible, if one looks beyond the slightly differently shaped face, to read the emotions of apes as easily and as accurately as one reads the emotions and feelings of other human beings. There are few feelings that apes do not share with us, except perhaps self-hatred. They certainly experience and express exuberance, joy, guilt, remorse, disdain, disbelief, awe, sadness, wonder, tenderness, loyalty, anger, distrust, and love. Someday, perhaps, we will be able to demonstrate the existence of such emotions at a neurological level. Until then, only those who live and interact with apes as closely as they do with members of their own species will be able to understand the immense depth of the behavioral similarities between ape and man.

Knowing what we feel is one way to judge whether an animal feels something similar, but may not be the only, or even the best way. Are animals' similarities or differences from humans the only, or even the most important, issue? Surely we can train ourselves to an empathic imaginative sympathy for another species. Taught what to look for in facial features, gestures, postures, behavior, we could learn to be more open and more sensitive. We need to exercise our imaginative faculties, stretch them beyond where they have already taken us, and observe things we have never been able to see before. We need not be limited by ourselves as the reference point, by what has already been written, by the existing consensus among scientists. What do we have to lose in taking the imaginative leap to broaden our sympathies and our horizons? I decided to explore what had been written about animals in scientific studies to see whether they contained buried information about their emotions, even if they did not contain explicit discussions of such matters. As yet no prominent scientist has undertaken a sustained treatment of animal emotions. It is to be hoped, for the sake of animals as well as humans, that scientists will be persuaded to look more seriously at the feelings of the animals who share the world with us.

In this book I try to show that animals of all kinds lead complex emotional lives. Although many scientists have believed that the animals they observed had emotions, few have written about it. This is why my co-author and I have sifted a large body of scientific literature, looking for the unacknowledged evidence. I have drawn on a long list of expert witnesses, in particular scientists who have studied wild animals in the field. I have kept largely to work by recognized scientists, so that even skeptics will see that evidence comes from a wide range of careful studies of animals in different environments.

These field studies show what most laypeople have always believed: that animals love and suffer, cry and laugh; their hearts rise up in anticipation and fall in despair. They are lonely, in love, disappointed, or curious; they look back with nostalgia and anticipate future happiness. They *feel*.

No one who has lived with an animal would deny this. But many scientists do just that, which is why I have tried to address their worries in more detail than might be necessary for the ordinary person. "It's obvious," says the pet owner; "It's an enormous claim," says the scientist. This book attempts to bridge the gap between the knowledge of the person who has always observed animals without prejudice, and the scientific mind that does not want to venture into such emotional territory.

Many scientists have avoided thinking about the feelings of animals because they have been frightened—and realistically so—of being accused of anthropomorphism. That is why I have looked carefully at the issue of anthropomorphism. If it can be disposed of as a false criticism, then the study of animal emotions can proceed on a scientific basis, freed from a bogus fear.

I have also tried to look objectively at the arguments of evolutionary biology and ask, when do they help explain the real emotional lives that animals display and when are they used to dismiss that reality?

As you read you may be surprised by the unexpected emotional behavior of some animals: an elephant who keeps a pet mouse; a chimpanzee awaiting the return of her dead baby; a bear lost in rapture as it watches the sunset; ice-skating buffalo; a parrot who means what he says; a dolphin inventing her own games—and through it all, scientists who refuse to acknowledge what will probably seem obvious to you.

In the conclusion I will discuss some of the moral choices that flow from an accurate understanding of animal emotions. We will have seen that animals feel anger, fear, love, joy, shame, compassion, and loneliness to a degree that you will not find outside the pages of fiction or fable. Perhaps this will affect not only the way you think about animals, but how you treat them. The clearer it became to me that animals have deep feelings, the more outraged I grew at the thought of any kind of animal experimentation. Can we justify these experiments when we know what animals feel as they undergo these tortures? Is it possible to go on eating animals when we know how they suffer? We are horrified when we read, even in fiction, of people who kill other people in order to sell parts of their bodies. But every day elephants are slaughtered for their tusks, rhinos for their horns, gorillas for their hands. My hope is that as it begins to dawn on people what feeling creatures these animals are, it will be increasingly difficult to justify these cruel acts.

JEFFREY MOUSSAIEFF MASSON
Half Moon Bay, April 1995

In Defense of Emotions

Somewhere in India, a blind river dolphin seeks her companion. Under the dark waters of the Ganges she will sleep next to him. She has never needed to see. These dolphins find everything they want and need by listening to echoes. Above them in the sky, two cranes from the east are flying back from China to their western breeding territory in Siberia. The cranes are a mile up in the sky, looking down with their golden eyes. What is in their hearts, or in the hearts of the dolphins? Wholly apart from us, their lives of turmoil and satisfaction are not beyond our imagination. When the dolphin rises out of the muddy waters, or the cranes stretch their necks in flight, we are filled with a sudden sense of familiarity, the recognition that we share an emotional heritage. They feel and we feel, no matter how difficult it is to know just what their feelings are.

After a promising start over 120 years ago, when Darwin explored the terrain in his book *The Expression of the Emotions in Man and Animals*, very few scientists have acknowledged, researched, or even speculated about animal emotions. So persistent are the forces that militate against even admitting the possibility of emotions in the lives of animals that the topic seems disreputable, almost taboo. The scholarly literature on animals contains many observations, accounts, and anecdotes that suggest emotions the animals may be experiencing or expressing, or at least call for further research into this possibility. Yet little to none is forthcoming.

G. G. Rushby, a game warden doing "elephant control work" in Tanzania (then Tanganyika), saw three female elephants and a half-grown male in tall grass. Since his job was to keep the elephant population down, he shot the females—and slightly wounded the half-grown animal. To his dismay, he suddenly saw two elephant calves, who had been with the females but had been hidden in the grass. He moved toward them, shouting and waving his hat, hoping to drive them back to the larger herd, where other elephants would adopt them. The wounded elephant was dazed and helpless and did not know which way to turn. Instead of fleeing, the orphaned calves pressed themselves against him and supported him, and led him away from danger.

Terror, compassion, bravery—accounts like this, systematically developed, could provide evidence for a world of deep emotional experience on the part of animals, but there appears little place for them in scientific literature. Onetime incidents are dismissed as "anecdotes," yet there is no reason to ignore rare events. And when it is feasible to collect other instances or even to repeat rare events, this is seldom done, so scathing do scientists find the charge of "using anecdotal evidence." Discussing the ability of two symbol-using chimpanzees, Sherman and Austin, to improvise unusual and impressive combinations of symbols, primate researcher Sue Savage-Rumbaugh calls such spontaneous occurrences "arguably the most important sort of data we have," yet notes, "we have avoided describing these in our published reports."

Without question, anecdotes do present hurdles for scientists, including the inability to control the circumstances around the event, the frequent lack of documentation, and the impossibility of generating statistics from a single occurrence. But even when an event is meticulously recorded and takes place in a controlled situation, like the symbol combinations used by Sherman and Austin, the onetime nature of the event precludes its use in many scientists' eyes. Experimental evidence is given almost exclusive credibility over personal experience to a degree that seems almost religious rather than logical.

Jane Goodall finds the scientific reluctance to accept anecdotal evidence a serious problem, one that colors all of science. "I've always collected anecdotes, because I think they're just terribly, terribly important—whereas most scientists scorn the anecdotal. 'Oh, that's merely anecdotal.' What is anecdotal? It's a careful description of an unusual event." She tells of a research assistant in a laboratory charged with logging the response of male rhesus monkeys to females, some of whom were being treated with hormones or had had their ovaries removed. "She told me . . . the most fascinating thing to her was that there was one old female that she observed in all these different states, ending with having her ovaries out, and whatever state she was in, she was the most popular. But she was *one* monkey, and that was ignored. There must be literally millions of observations like that that have never crept into the literature." Such observations would provide a rich and suggestive ground for analysis and further investigation, yet there are almost none. While it is possible and customary to describe such events without using words that connote emotion, such a lean description is not necessarily more accurate.

This book defines *emotions* as subjective experiences, as what people refer to when they say "I feel sad," or "I am happy," or "I am disappointed," or "I miss my children." An emotion is not distinguished from a feeling, a passion, a sentiment, or what scientists call "affect." *Mood* refers to a feeling that lasts for some protracted time. These words refer simply to inner feeling states, to something that is felt.

The Practical Impossibility of Ignoring Emotion

Most people who work closely with animals, such as animal trainers, take it as a matter of fact that animals have emotions. Accounts by those who work with elephants, for example, make it clear that one ignores an elephant's "mood" at one's peril. The British philosopher Mary Midgley puts it well:

> Obviously the mahouts may have many beliefs about the elephants which are false because they are "anthropomorphic"— that is, they misinterpret some outlying aspects of elephant behaviour by relying on a human pattern which is inappropriate. But if they were doing this about the basic everyday feelings—about whether their elephant is pleased, annoyed, frightened, excited, tired, sore, suspicious or angry—they would not only be out of business, they would often simply be dead.

Training an animal will meet with little success if the trainer has no insight into the animal's feelings. Some trainers say they work with certain animals better than others because they understand the feelings of that species or individual better. Circus trainer Gunther Gebel-Williams noted individual differences in the emotions of the tigers with whom he worked: "Not every tiger . . . can be trained to jump through a ring of fire. When I incorporated that trick into the tiger act I had to find several from among the twenty I was working with at the time who were not afraid of fire. That was no easy task, because most tigers will not go near flames."

The fear of committing anthropomorphism can handicap an animal trainer, says Mike Del Ross, training supervisor at Guide Dogs for the Blind in San Rafael, California: "The more you open yourself up to trying to read the dog, the more you concentrate and read the dog better."

Asked whether they would still want to work with dogs if dogs had no emotions, trainers were startled by the very idea. Kathy Finger replied, "Probably not, because I think reading emotions is part of dealing with dogs—loving them, respecting them." Del Ross exclaimed, "No way. What would there be if they didn't have emotions?"

Such empathy in the direct scientific observation of animals is controversial. But wondering what one would feel in the place of an animal can be fruitful. Most scientists working with animals in the wild make inferences based on empathy, to make sense of their behavior, such as, "If I had just lost my closest companion, I, too, would not feel like eating for some time." Thinking about feelings has proven to be a valuable way of thinking about behavior.

The Emotions of Captivity—"That Doesn't Count"

Evidence of emotion in captive animals and pets is often discounted as irrelevant. Captive animals, the argument goes, are in unnatural situations, and what domesticated animals do is irrelevant to what animals are *really* like, as if they are not really animals. While genuinely domesticated animals *are* different from wild animals, *domestic* and *tame* do not mean the same thing. Domestic animals are animals that have been bred to live with humans—they have been changed genetically. Dogs, cats, and cows are domestic animals. Captive animals like elephants are not, since through the generations that people have trained elephants, they have almost invariably caught and tamed wild elephants, rather than *bred* elephants. Since the nature of elephants remains unchanged, observations on tame or captive elephants are in fact highly relevant to free-living elephants.

While domesticated animals and wild animals may not be the same, they still have much in common; information about one can be relevant to the other. As field biologist George Schaller has written, "A loving dog-owner can tell you more about animal awareness than some laboratory behaviorists." Biologist Lory Frame studied wild dogs in the Serengeti and made the intriguing observation that the dominant animals (the only ones in a pack who usually breed), seemed much less like domestic dogs. "Intuitively I seemed to understand Maya and Apache. And that, I realized, was because their subordinate behavior reminded me of domestic dogs. Not that my family's dog was the cringing sort—on the contrary, he bullied me when I was a child. But Maya's tail-wagging manner was reminiscent of the behavior people expect, and usually get, from their pet dogs. Dominant wild dogs, however, are something else. . . . I seldom saw one grin or wag his tail. Serious they seemed, and dangerous. If I met Sioux on foot I'd get up the nearest tree. With Maya I'd be more likely to pat her on the head and offer her a biscuit." What she knew from her experience with domesticated dogs helped inform her observations of wild dogs.

That captive and domestic animals are in "an unnatural situation" is simply not a valid reason to treat observations of them less seriously. Humans are in just as unnatural a situation. We did not evolve in the world in which we now live either, with its deferred rewards and strange demands (sitting in classrooms or punching time clocks). All the same, we do not dismiss our emotions as not existing or inauthentic simply because they don't take place in small groups of hunter-gatherers on an African savanna, where human life is thought to have begun. We are ourselves domesticated animals. We can be at a distance from our "origins" and still claim that our emotions are real and characteristic of our species. Why can't the same be true of animals? It is not natural for humans to be in prison. Yet if we are put in prison and feel emotions that we don't usually feel, no one doubts that they are real emotions. An animal in a zoo, or kept as a pet, may feel emotions that it would not otherwise have felt, but these are no less real.

To find out if her observations of captive dwarf mongooses told her anything accurate about mongooses in a natural state, Anne Rasa, author of *Mongoose Watch*, went to Kenya to study them in the bush for several years. She discovered that the behavior of captive mongoose groups in large enclosures closely followed that of wild ones with two exceptions. The wild mongooses had to spend much more time gathering food, and hence less time playing and socializing. Their lives were also strongly colored by the actions of other species. Eagles and snakes preyed on them, so they spent a significant time mobbing snakes to drive them away. They quarreled with the larger blacktip mongooses. They usually ignored lizards and ground squirrels, but occasionally tried to play with them. In other words, their emotional range was to some extent determined by the opportunities that presented themselves, but curiosity and play were common to both captive mongooses and those in a natural state.

On the other hand, the conditions of captivity can certainly change the way animals behave. Female baboons kept together in a cage form a rigid hierarchy unlike anything seen in the species in the wild. The point is not that captivity never changes emotions and behavior, only that both captive and wild animals appear to have feelings, and that the emotions of captive animals are as real as those of wild animals, and therefore equally worthy of study.

Complexity of Emotion

Emotions seldom come pure, in isolation from other emotions. In people, anger and fear, fear and love, love and shame, shame and sorrow often converge in particular situations. Animals may also experience a mix of emotions. Perhaps a dolphin mother who carries her dead baby around with her for several days feels both love and sorrow. Hope Ryden describes a half-grown elk calf guarding the body of another calf killed by coyotes after the elk herd had moved on. For at least two days the calf straddled the body, aggressively chased coyotes away, and from time to time sniffed and nuzzled the face of the dead calf. Eventually (after coyotes had succeeded in partly eating the body) the calf moved on. The calf may have felt grief; it may have felt lonely for the rest of the herd; it may have felt anger at the coyotes; it may have feared the coyotes. Perhaps it felt love for the dead calf. That feelings may be complex and multifaceted or difficult to interpret does not mean they do not exist.

Animals do not all have the same emotions, any more than humans do. Just as the behavior of animal species differs, their feeling lives may differ as well. This is often overlooked when people argue from animal examples. "Geese mate for life," people declare. Or, "Robins kick their young out of the nest when they're old enough to be on their own." "The dog doesn't stay and help the bitch raise the puppies—that's just the way it is." This wrongly assumes that all animals are the same and therefore we can draw the same conclusions about people. But while geese mate for life, grouse do not. The male grouse mates with as many females as he can and leaves them to raise the young on their own. The female Tasmanian native hen often mates with two males and the trio raises the young together. While robins fledge at an early age, condors stay with their parents for years. Male and female wolves raise their puppies together. These differences often produce a kind of sociobiology parlor game in which people try to prove points about human behavior by pointing to an animal species that exhibits the behavior they want to define as "natural" for humans. But animal species may also differ in the content of their emotions. Evidence that elephants feel compassion or sorrow does not mean that hippos feel compassion or that penguins feel sorrow. Perhaps they do, perhaps not.

Animals also differ from one another as individuals. Among elephants, for example, one may be timid and another bold. One may be prone to attacks of rage, another peaceable. One Victorian commented on working elephants in Rangoon, "There are willing workers and there are skulkers; there are gentle tempers, and there are others as dour as a door-nail. Some of them will drag a log two tons in weight without a groan; while others, who are equally powerful but less willing, will make a dreadful fuss over a stick that is, comparatively speaking, nothing." Of a species he hunted, Theodore Roosevelt wrote: "[B]ears differ individually in courage and ferocity precisely as men do. . . . One grizzly can scarcely be bullied into resistance; the next may fight to the end, against any odds, without flinching, or even attack unprovoked. . . . Even old hunters—who indeed, as a class, are very narrow-minded and opinionated—often generalize just as rashly as beginners."

Views of Animal Emotion; Lay and Scientific

Most ordinary people who have direct contact with animals freely concede the reality of animal emotions. Their belief arises from the evidence of their senses and logical deduction. A person who hears birds attacking a cat near their nest usually experiences them as angry. When we see a squirrel flee from us, we think that it is afraid. We see a cat licking its kittens and feel it loves them. We see a bird throbbing with song and suppose it to be happy. Even those with only indirect experience of animals often recognize what they see to be an emotional state, a feeling, which they correlate to a similar human feeling. In this respect the layperson's description of animal life may be more accurate and is certainly richer than the standard behaviorist's description, which shows no effort to investigate animal emotions systematically or in depth.

Despite the lack of sustained scholarly work on animal emotions, there is today a greater interest in the realities of the lives of animals than ever before. Practitioners in a wide range of disciplines share an increasing awareness of the complexity of animal actions—cognitive, perceptual, and behavioral, individual and social—and correspondingly greater humility in the face of questions of animal capacities. Humans are no longer as prepared to pronounce upon what an animal can and cannot be and do. We are starting to be clear that we do not know and are only beginning to learn.

While the study of emotion is a respectable field, those who work in it are usually academic psychologists who confine their studies to human emotions. The standard reference work, *The Oxford Companion to Animal Behavior*, advises animal behaviorists that "[O]ne is well advised to study the behaviour, rather than attempting to get at any underlying emotion." Why? They may be elusive or difficult to measure, but this does not mean animal feelings do not exist and are not important.

Human beings are not always aware of what they are feeling. Like animals, they may not be able to put their feelings into words. This does not mean they have no feelings. Sigmund Freud once speculated that a man could be in love with a woman for six years and not know it until many years later. Such a man, with all the goodwill in the world, could not have verbalized what he did not know. He had the feelings, but he did not know about them. It may sound like a paradox—paradoxical because when we think of a feeling, we think of something that we are consciously aware of feeling. As Freud put it in his 1915 article "The Unconscious": "It is surely of the essence of an emotion that we should be aware of it." Yet it is beyond question that we can "have" feelings that we do not know about.

Psychiatric lexicons contain the term *alexithymia* for the condition of certain people who cannot describe or recognize emotions, who are able to define them "only in terms of somatic sensations or of behavioral reaction rather than relating them to accompanying thoughts." Such people are handicapped by their inability to understand what feelings are. It is curious that the study of animal behavior should demand that its practitioners turn themselves into alexithymics.

Defining the Emotions

Psychological theorists speak of a set of fundamental human emotions that are universal, discrete, and which they consider innate. These fundamental emotions are like the primary colors and can give rise to many variations. One psychologist compiled a list of 154 emotion names, from abhorrence to worry. Theorists do not agree on which emotions are the basic ones. René Descartes said there were six basic emotions: love, hate, astonishment, desire, joy, and sorrow. Immanuel Kant found five: love, hope, modesty, joy, and sorrow. William James defined four: love, fear, grief, and rage. Behaviorist J. B. Watson postulated three basic emotions, X, Y, and Z, roughly equivalent to fear, anger, and love. Modern theorists like Robert Plutchik, Carroll Izard, and Silvan Tomkins found either six or eight basic emotions—but not the same ones. On most modern lists *love* is not included as an emotion. Many scientists prefer to call it a drive or a motivation, if they refer to it at all. All the emotions in these commonly used and accepted lists have been thought by some researchers to be observed among animals.

In addition to these, there are probably other emotions and variations within them that from time to time everybody, from whatever culture, feels. Compiling a full list can be hazardous, however, as the Polish linguist Anna Wierzbicka points out when she observes that in some non-Western cultures, for example in Aboriginal Australia, a concept related but not identical to shame plays a social role evidently missing in our culture. The word describing this emotion can include the English concept of "shame," "embarrassment," "shyness" and "respect." Yet it seems likely that the feeling itself would be recognizable at least approximately to somebody from another culture.

We should be wary of confining any emotion to only one part of the world. After all, it was not so long ago that ethnologists thought that there were some cultures (obviously inferior) where the full range of Western emotions could not be expressed, and thus were probably not experienced. It seemed then as pointless to inquire into compassion or aesthetic awe among certain hill tribes as it now appears to catalog aesthetic rapture among bears. One of the "great" anthropological texts of the beginning of this century was titled *Mental Function in Inferior Societies,* written by L. Lévy-Bruhl, who was professor at the Sorbonne. Such prejudice is slowly receding. The capacity to feel all emotions may be universal. Great literature suggests that certain feeling states are universal, or at least that the capacity to experience them crosses cultures, although different cultures and different individuals may describe them differently, or attach differing importance to subtleties of feeling. If feelings can cross cultures, it seems likely they can cross species.

This book discusses animal emotions following the order in which people find them plausible. Humans are most ready to consider the possibility of other animals having the emotion of fear. Love, sorrow, and joy are considered "nobler," hence less likely to be granted to others, especially animals. Although many people are very ready to speak of anger in animals, some experienced animal trainers argue that animals do not feel this emotion. The sociobiological debate over altruism has resulted in a widespread denial of the possibility of compassion in animals. As for shame, a feeling for beauty, creativity, a sense of justice, and other even more elusive capacities: these are the least likely to be ascribed to animals.

The Functions and Benefits of Emotion

What are feelings for? Most nonscientists will find this a strange question. Feelings just are. They justify themselves. Emotions give meaning and depth to life. They need serve no other purpose in order to exist. On the other hand, many evolutionary biologists, in contrast to animal behaviorists, acknowledge some emotions primarily for their survival function. For both animals and humans, fear motivates the avoidance of danger, love is necessary to care for young, anger prepares one to hold ground. But the fact that a behavior functions to serve survival need not mean that that is why it is done. Other scientists have attributed the same behavior to conditioning, to learned responses. Certainly reflexes and fixed action patterns can occur without feeling or conscious thought. A gull chick pecks at a red spot above it. The parent has a red spot on its bill; the chick pecks the parent's bill. The gull parent feeds its chick when pecked on the bill. The baby gets fed. The interaction need have no emotional content.

At the same time, there is no reason why such actions cannot have emotional content. In mammals—including humans—that have given birth, milk is often released automatically when a new baby cries. This is not under voluntary control; it is reflex. Yet this does not mean that feeding a new baby is exclusively reflex and expresses no feelings like love. Humans have feelings about their behavior even if it is conditioned or reflexive. Yet since reflexes exist, and conditioned behavior is widespread, measurable, and observable, most scientists try to explain animal behavior using only these concepts. It is simpler.

Those who argue against speaking of emotion and consciousness in animals often appeal to the principle of parsimony, or Ockham's razor. This principle holds that one should choose the simplest explanation for a phenomenon. Animal behaviorist Lloyd Morgan's version reads: "In no case may we interpret an action as the outcome of an exercise of a higher physical faculty, if it can be interpreted as the outcome of the exercise of one which stands lower in the psychological scale." This rule of giving credence to only the lowest or simplest explanation for behavior is not unassailable. Many questionable assumptions lie buried in the assessment of faculties as higher and lower. Emotions are typically considered higher faculties for no very clear reason. Moreover, the world is not necessarily a parsimonious place. As Gordon Burghardt has pointed out, "The origin of life by creation is simpler than the indirect methods of evolution."

Preferring to explain behavior in ways that fit science's methods most easily, many scientists have refused to consider any causes for animal behavior other than reflexive and conditioned ones. Scientific orthodoxy holds that what cannot be readily measured or tested cannot exist, or is unworthy of serious attention. But emotional explanations for animal behavior need not be impossibly complex or untestable. They are just more difficult for the scientific method to verify in the usual ways. Cleverer and more sophisticated approaches are called for. Most branches of science are more willing to make successive approximations to what may prove ultimately unknowable, rather than ignoring it altogether.

Funktionslust

Evolutionary biology offers further support for the view that animals feel. In this model, anything that enhances survival has selective value. Emotions can motivate survival behavior. An animal who is afraid of danger and runs away may survive over the one who does not, while another animal that angrily defends its territory may live longer and better. An animal that loves and protects its offspring may leave more descendants. An animal may take pleasure in the ability to run swiftly, fly strongly, or burrow deeply. The old German term *funktionslust* refers to pleasure taken in what one can do best—the pleasure a cat takes in climbing trees, or monkeys take in swinging from branch to branch. This pleasure, this happiness, may increase an animal's tendency to do these things, and will also increase the likelihood of its survival.

But not all actions driven by emotion have survival value. A loving animal may leave more offspring, thus making love an aid to survival; but a loving animal may also care for disabled offspring or companions that have no chance of surviving, or expose itself to hazards mourning dead ones. It may adopt the babies of others, not passing on its own genes. These actions would not enhance, and would probably decrease, its own fitness. Perhaps animals take certain actions because of what they feel, not simply because of any survival advantage conferred. Yet lovingness could still have survival value, because the net effect would be to leave more offspring. If a behavior that is usually adaptive also occurs when it has no survival advantage, this may mean that an overarching emotion, not a narrow adaptation, drives the behavior. Systematic observations of this sort could promote theorizing on emotions, even test their existence. If a usually adaptive behavior occurs in an unadaptive situation, an overarching emotion, not a narrow adaptation, may drive the behavior.

Biologists often point to a behavior's evolutionary advantage as a way to sidestep the question of emotions. Scientists sometimes argue that the songbird is not singing with joy, nor singing because he finds his song beautiful, but because he is establishing territory and advertising his fitness to possible mates. Thus to view birdsong as an aggressive and sexual act provides a genetic explanation for the behavior. The bird's song may announce his territorial claims, and may indeed attract a mate, but that does not preclude the bird singing because he is happy and finds his song beautiful. As primatologist Frans de Waal points out, "When I see a pair of parrots tenderly and patiently preening each other, my first thought is not that they are doing this to help the survival of their genes. This is a misleading manner of speaking, as it employs the present tense, whereas evolutionary explanations can deal only with the past." Instead de Waal views the birds as expressing love and expectation, or, retreating a little, "an exclusive bond."

Similarly, human behavior that *can* be viewed as increasing survival fitness often cannot be explained from that standpoint only, as sociobiologists sometimes attempt to do. When monogamous humans have affairs, they are not generally thinking about maximizing reproductive chances by impregnating females other than the one with whom a substantial parental investment is being made, or about mating with genetically superior males for the ben-

efit of their progeny. Indeed, adulterers usually try to avoid reproduction. Sexual abuse of children has no survival value either, yet is
common. If humans are subject to evolution but have feelings that
are inexplicable in survival terms, if they are prone to emotions
that do not seem to confer any advantage, why should we suppose
that animals act on genetic investment alone?

A Double Standard

As human beings, we clearly apply different standards to ourselves than to other animals. Humans are conceded to have emotions. The usual reason given is that feelings are expressed in language, using words like "I love you," or "I don't care," or "I am
sad." People live much of their lives according to expressions of
feelings in themselves or in others. Although it is widely agreed
that some people lie about their feelings to gain an advantage, and
some people make mistakes about their feelings, or do not know
what they really feel, or express them without credibility, few
doubt that feelings exist—one's own, and those of others. The
primary method of reasoning seems to be analogy and empathy: we
know we have feelings because we feel moved by them, and others
do and express similar things, so we believe that they have feelings
too.

Such reasoning has its limitations. We learn from personal
experience that other humans can feel gratitude because they say so
and act as though they do. By itself, this sheds no light on whether
a lion can feel gratitude. On the other hand, humans, even when
embedded in sophisticated cultural environments, remain very
much a species of animal; the relation of the physical to the psychic
ingredients of emotions may well be shared. While emotions cannot be reduced simply to a blend of hormones, to whatever extent
hormones contribute to emotional states in humans, they probably
also do so in animals. Substances like oxytocin, epinephrine, sero-
tonin, and testosterone—all of which are thought to affect human
actions and feelings—are found in animals as well. Grossly oversimplified explanations of human behavior in terms of hormones

have proved not only faulty but pernicious; care should be taken to avoid the same mistake in explaining animal behavior.

Belying the closely held belief that emotions are the exclusively human products of our unparalleled mental powers, the physical pathways of human emotion are among the most primitive. The part of the brain called the limbic system, which is thought to mediate emotion, is one of the most phylogenetically ancient parts of the human brain, so much so that it is sometimes called "the reptile brain." From a purely physical standpoint, it would be a biological miracle if humans were the only animals to feel. Can it then be shown, say, that a cat loves her kittens or that kittens love their mother? If measurements showed hormone levels surging in the cat's bloodstream when she sees her kittens, and electrical activity spiking in certain parts of the cat's brain, would that be accepted as proof? Many scientists would still say no; we can never know if a cat loves. Yet most observers already believe that the cat loves the kittens, simply on the basis of her behavior. Scientists prefer not to say so.

Could it be that the statement "The ape is clearly sad" is not so different from "John is clearly sad"? The *clearly* signals an interpretation; it refers to clues that are socially agreed upon to indicate sadness. John is staring at the ground for hours and sighing. So is the ape. John may refuse to eat. So might the ape. John refuses to speak; when asked how he feels, he stares past the speaker. We do not, for that reason, say that he cannot feel sorrow or he would say so. We can be wrong about the ape. We can also be wrong about John. John might, in fact, be feeling something entirely different— apathy, perhaps, or existential despair. We may have misunderstood his actions, his facial expressions, and his vocalizations. *Clearly* is a statement about the kind of evidence we think we have, but our evidence may not be as good for people, nor as poor for other animals, as we have assumed.

The Slippery Clues of Language

Humans do have the advantage of language, one of the biggest differences between humans and other animals. Animals cannot speak of their feelings in a way humans can reliably understand, although the language barrier between humans and animals is not absolute. But language is not entirely trustworthy as a yardstick of feeling between humans. Verbal assertion of a feeling does not prove the emotion exists, nor does the inability to verbalize an emotion prove it does not. Some profoundly retarded humans cannot speak their feelings; this does not mean they do not have them. Mute humans feel. Intellectually sophisticated people can lie about their feelings or conceal them. Intellectual capacity may distinguish people from other animals, even if only in degree, but even among humans, intelligence and emotion are not closely correlated.

Language is a part of culture, and cultures around the world seem to make many of the same distinctions between emotions and to refer to similar experiences. But can we feel an emotion for which our culture provides no word or no examples? No doubt there are emotions promoted in one culture and not another, but this does not mean they are not experienced in all of them. It may be difficult to define or express them, given the language to which one is born; it may even be difficult to think about them, and especially to convey them to another person. Yet the feelings themselves may have a certain autonomy such that they may nonetheless be felt. Similarly, animals may have emotional experiences it would be hard to express or put into words, even if they had the capacity to use words, but they would not for that reason cease to be real feelings. The language barrier notwithstanding, humans may well share with animals the vast majority of feelings of which they are capable.

The prejudice has long existed that only humans think and feel because only humans can communicate thoughts and feelings in words, whether written or spoken. Descartes, the seventeenth-century French philosopher, believed animals to be "thoughtless brutes," *automata*, machines:

There are [no men] so depraved and stupid, without even excepting idiots, that they cannot arrange different words together, forming of them a statement by which they make known their thoughts; while, on the other hand, there is no other animal, however perfect and fortunately circumstanced it may be, which can do the same . . . the reason why animals do not speak as we do is not that they lack the organs but that they have no thoughts.

An unknown contemporary of Descartes put this position starkly:

The [Cartesian] scientists administered beatings to dogs with perfect indifference and made fun of those who pitied the creatures as if they felt pain. They said the animals were clocks; that the cries they emitted when struck were only the noise of a little spring that had been touched, but that the whole body was without feeling. They nailed the poor animals up on boards by their four paws to vivisect them to see the circulation of the blood, which was a great subject of controversy.

Voltaire responded that, on the contrary, vivisection showed that the dog has the same *organes de sentiment* that a human has. "Answer me, you who believe that animals are only machines," he wrote. "Has nature arranged for this animal to have all the machinery of feelings only in order for it not to have any at all?" Elsewhere, in *Le philosophe ignorant*, he criticizes Descartes by saying that he "dared to say that animals are pure machines who looked for food when they had no appetite, who had the organs for feeling only to never have the slightest feeling, who screamed without pain, who showed their pleasure without joy, who possessed a brain only to have in it not even the slightest idea, and who were in this way a perpetual contradiction of nature." As early as 1738, Voltaire talked about the humane feelings of the great English physicist Isaac Newton and how, like the philosopher John Locke, he was convinced that animals had the same sentiments man did. Voltaire writes: "He [Newton] believed that it was a very terrible contradiction to believe that animals could feel, and yet cause them to suffer."

It is true that most animals have no speech that humans yet understand. But is the absence of speech, after all, as important an indication of feelings as some philosophers have imagined it to be? Several chimpanzees and other great apes have American Sign Language (ASL) vocabularies of more than a hundred words. They communicate not only with humans but with members of their own species. Would it not be parsimonious to suppose that they had previously communicated some of these same thoughts to other apes via means other than human sign language? Why would they wait for scientists before doing something they were already capable of doing? The fact that apes do not have human vocal cords does not mean that they must remain uncommunicative. Following a first flush of excitement, the overwhelming response of the scientific community to signing apes has been to ignore or disbelieve them, both as individuals and as a species. Given that statements made by apes about food and toys are attacked, one can only imagine the reaction to statements about their feelings. Rooted prejudice claims that animal feelings cannot be known because animals cannot speak; when they do speak in a human tongue, the claim is that what they are saying cannot possibly mean what humans mean.

Even when animals speak our language, humans do not always take them at their word. For sixteen years Alex, an African grey parrot, has been trained by psychologist Irene Pepperberg, who researches the bird's cognitive abilities. Alex is one of the few parrots in the world who has been demonstrated to understand the meaning of the words he speaks. He knows the names of fifty objects, seven colors, and five shapes. He can enumerate up to six objects and say which of two objects is smaller. Alex has also picked up many "functional" phrases. He has learned "I'm gonna go now," something he hears people say in Pepperberg's laboratory. Pepperberg describes how, when Alex is scolded, "We say, 'No! Bad boy!' We walk out. And he knows what to say contextually, applicably. He brings us back in by saying, 'Come here! I'm sorry!' " Alex learned to say he was sorry by hearing humans say it.

He knows when to say it. Does he feel regret? "He bites, he says, 'I'm sorry' and he bites again," says Pepperberg, somewhat irritably. "There's *no* contrition!" Just like many people.

Here is an animal who appears to be verbally reporting an emotional state—regret—but we don't believe him. If he were really sorry (as we understand the term) for biting, would he immediately bite again? Perhaps he would. Whatever is going on inside Alex, he is motivated enough to learn human words for human feelings—possibly to make humans into more satisfactory parrot companions. Alex may not feel contrition about hurting someone. Pepperberg may have no word for what Alex wants from her either; she may never have felt what Alex feels. Humans are surprisingly deficient in vocabulary for positive social emotions, and unduly successful at naming negative individualistic ones. Could there not be gradations of social proximity and affection at the top of the forest canopy for which humans are functionally emotional illiterates? Maybe we have something to learn.

Communication Without Language

Nonverbal communication among humans has sparked increasing interest among academics and therapists in the last few years. Many complex mental states are conveyed more conveniently by gestures than by sentences, while others appear to escape verbal language entirely. Attempts to convey subtle or elusive feelings leave everybody with a sense of the inadequacy of speech. Poetry, after all, is an attempt to convey feelings, moods, states, and even thoughts that are hard to grasp and that seem to defy language in prose. And some feelings do in fact elude language, even poetry, altogether. The fine arts and silence pick up where words leave off.

There is little doubt that humans communicate thoughts and feelings *without* words; indeed, there is growing evidence that a great part of communication with others takes place outside verbal speech. Just as humans communicate through body language, gestures, and expressive acts, formalized through mime and dance, consideration should be given to the nonverbal statements about feelings that animals make.

Animals communicate information through posture, vocalizations, gestures, and actions, both to other animals and to humans who are attentive. Although study of these patterns is improving, even specialists can be rather poor at interpreting this information; this is especially true for those unfamiliar with the species. The animals themselves are much better at understanding these signals, even across species. Indeed, Elizabeth Marshall Thomas speculates that animals are much better at reading human body signals than humans are at reading animal signals of any kind. "Our kind may be able to bully other species not because we are good at communication but because we aren't." De Waal complains that apes are so good at reading human body language as to leave people who work with them feeling transparent.

After fifteen years of studying red foxes, raising them and living with them, David Macdonald understands their body language. He can tell at a glance a happy fox, an excited fox, a nervous fox. He freely writes of them as playful, furious, besotted, fearful, confident, contented, flirtatious, or humiliated. His *Running with the Fox* illustrates fox body language so that those less familiar with foxes can figure it out. Yet because the emotions of animals are not scientifically respectable, when Macdonald discusses whether foxes enjoy killing, he retreats, with the caveat, "assuming they are subject to emotions recognizable to humans . . ." He calls this question "philosophically unanswerable." But to most lay people it is no more philosophically unanswerable than the question of whether other humans have emotions, including sadism.

In Konrad Lorenz's *The Year of the Greylag Goose*, the caption to one photograph of a gander reads: "After Ado [another gander] had appropriated Selma [his former mate], Gurnemanz went to pieces, as can be seen in this picture." To a person only casually familiar with geese, this cannot be seen at all. It might as easily be a happy goose or a furious goose. A goose does not have a mobile face, so there is little by way of facial expression. Lorenz, from long experience, knows a goose's body language and can read it. Gurnemanz's posture and neck position tell of his submission and demoralization. Elsewhere Lorenz describes goose postures, gestures, and sounds as victorious, uncertain, tense, glad, sad, alert, relaxed, or threatening.

The point is that a goose or other animal may be a quivering mass of emotion. Its feelings may be "written all over its face," and it may only take practice to read that writing. We are restricted only by ignorance, lack of interest, desire for exploitation (like wanting to eat them), or by anthropocentric prejudices that preclude us, as if by divine fiat, from recognizing commonality where it might exist. How can we be gods if animals are like us?

Exploring the Forbidden Subject

The standards for defining the existence of emotions in animals begin with those in common use for humans. One should demand no more proof that an animal feels an emotion than would be demanded of a human—and, like humans, the animal should be permitted to speak its own emotional language, which it is up to the beholder to understand.

Human emotions, too, escape exact scientific scrutiny. There is, in fact, no universally accepted scientific proof of human feelings. What one person *feels* is never entirely available to another. Not only is it uncertain our feelings are communicable; whether or not anyone understands the landscape of anyone else's inner life is ultimately unknowable. We think we know that people are sad, or lonely, or joyous, but it is hard to know the particularity of the accompanying mood. We may not be locked away in private universes of feeling, but another person's inner life, to the extent that it is individual, remains ultimately mysterious.

A history of human affairs in which fear, anger, love, pride, and guilt played no part would be strangely inadequate. Biographies without grief, sadness, and nostalgia would appear unreal. An ordinary person's life in which no one loves, is loved, or wants to be loved; in which no one fears anything; in which no one becomes angry or makes anyone else angry; in which the depths of despair remain unfathomed; in which no one feels pride in anything they do; in which no one is ever ashamed to do anything or feels guilty if they do—this would be an unnatural, unrealistic, paltry description. It would be neither believable nor accurate. It would be called inhuman. To describe the lives of animals without including their emotions may be just as inaccurate, just as superficial and distorted, and may strip them of their wholeness just as profoundly. To understand animals, it is essential to understand what they feel.

Unfeeling Brutes 2

Humans have historically been much concerned with distinguishing ourselves from beasts. We speak; we reason; we imagine; we anticipate; we worship, we laugh. They do not. The historical insistence on an unbridgeable gap between humans and other animals suggests that it serves some need or function. Why do we humans so frequently define ourselves by distinction from animals? Why should the distinction between man and beast matter?

Attempts to make this distinction fall largely into two categories. First, many cite human failings as unique, chief among which is fighting among ourselves. In these cases, the writer is usually trying to inspire his readers with moral resolutions. In the first century A.D., Pliny the Elder, in his *Natural History*, admonishes: "Lions do not fight with one another; serpents do not attack serpents, nor do the wild monsters of the deep rage against their like. But most of the calamities of man are caused by his fellow men." When, in 1532, Ludovico Ariosto in *Orlando Furioso* says, "Man is the only animal who injures his mate," this, too, is meant as an admonition. James Froude in his *Oceana* of 1886 claimed, "Wild animals never kill for sport. Man is the only one for whom the torture and death of his fellow creatures is amusing in itself." And even William James, in this century, wrote that "Man . . . is simply the most formidable of all the beasts of prey, and, indeed, the only one that preys systematically on his own species." In these examples, animals are not so much being observed as men are being exhorted to cease killing (usually) other men. They are intended to shame men into recognizing that they behave worse than animals.

The other—by far larger—category of man-beast contrasts cites human advantages: our intelligence, our culture, our sense of humor, our knowledge of death. In the nineteenth century William Hazlitt maintained, "Man is the only animal that laughs and weeps; for he is the only animal that is struck with the difference between what things are and what they ought to be." And in our century, the philosopher William Ernest Hocking claimed, "Man is the only animal that contemplates death, and also the only animal that shows any sign of doubt of its finality." Uniqueness is claimed for the human sense of humor, the ability to understand virtue, the ability to make and use tools. Again the authors seem more interested in making a didactic point for humans than in observing or understanding animals.

Human-animal comparisons have historically served as a rich source of moral instruction for humanistic philosophers, particularly during periods when the natural world was sentimentalized and viewed as a model. The most poetic was Buffon, the great nineteenth century French naturalist, who began his essay "On the Nature of Animals" by saying that animals cannot think or remember, but have feelings "to an even greater degree than humans do." Buffon believed there to be an advantage to an animal's purely feeling life. Humans, he wrote, lead lives of quiet desperation, and "most men die of sorrow." In contrast, "Animals do not search for pleasures where none can be found; guided by their feelings alone, they never make a mistake in their choice; their desires are always proportional to their capacity to enjoy; they feel as much as they enjoy and enjoy only as much as they feel. Man, on the other hand, wanting to invent pleasures, does nothing but spoil nature; wanting to force feelings, he only abuses his being, and digs a hole in his heart which nothing is capable of later filling." He ends by speaking of "the infinite distance that the Supreme Being has put between animals and [Man]."

Contemporary renditions of this contrast have been scarcely more grounded in animal reality and have not shed much more light on animals—or humans. Recently N. K. Humphrey wrote that "human beings have evolved to be the most highly social creatures the world has ever seen. Their social relationships have a depth, a complexity, and a biological importance to them, which no other animals' relationships come near." Considering how little is known about "other animals' relationships," this seems unwarranted.

How little we know, and how much we pretend to know, is illustrated by the fact that, until very recently, it was a canon of animal behavior that, of females, only the human experienced orgasm. As recently as 1979, anthropologist Donald Symons pronounced that the "female orgasm is a characteristic essentially restricted to our own species." When the question was actually investigated in the stump-tailed macaque, using the same physiological criteria used for humans, it was found that the female macaques did appear to experience orgasm. Primatologist Frans de Waal observes the same of the female bonobo (pygmy chimpanzee), from behavioral evidence. Like many questions specifically involving the human female, the truth is that not many scientists had ever considered the question systematically, let alone done the necessary field observation studies to find an answer. Perhaps it pleased most male scientists to imagine that while animal females sought sex only during an estrus cycle, and hence had sex only for reproduction, human females, due to their unique orgasmic capacity, wanted sex all the time.

Our Noble Feelings

People have always exalted certain "higher" feelings that are claimed to single us out among animals. Only humans, it is said, feel noble emotions such as compassion, true love, altruism, pity, mercy, reverence, honor, and modesty. On the other hand, people have often attributed so-called negative or "low" emotions to animals: cruelty, pride, greed, rage, vanity, and hatred. At play here appears to be a seemingly unbearable injury to our sense of uniqueness, to our entitlement to the special nobility of our emotional life. Thus not only whether animals can feel, but what they feel, is used to strengthen the species barrier. What lies behind this "us/them" mentality—the urge to define ourselves by proving we are not only different, but utterly different, including emotionally? Why should this distinction between man and beast be so important to humans?

A look at the distinctions humans draw among ourselves may provide a partial answer. Dominant human groups have long defined themselves as superior by distinguishing themselves from groups they are subordinating. Thus whites define blacks in part by differing melanin content of the skin; men are distinguished from women by primary and secondary sex characteristics. These empirical distinctions are then used to make it appear that it is the distinctions themselves, not their social consequences, that are responsible for the social dominance of one group over the other. Thus the distinction between man and beast has served to keep man on top. People define themselves as distinct from animals, or similar when convenient or entertaining, in order to keep themselves dominant over them. Human beings presumably benefit from treating animals the way they do—hurting them, jailing them, exploiting their labor, eating their bodies, gaping at them, and even owning them as signs of social status. Any human being who has a choice does not want to be treated like this.

A blatant example of many of these prejudices, with a suggestion of some of their social consequences, can be found in the article on "Animals" in the *Encyclopedia of Religion and Ethics*, written in 1908:

> Civilization, or perhaps rather education, has brought with it a sense of the great gulf that exists between man and the lower animals. . . . In the lower stages of culture, whether they be found in races which are, as a whole, below the European

level, or in the uncultured portion of civilized communities, the distinction between men and animals is not adequately, if at all, recognized. . . . The savage . . . attributes to the animal a vastly more complex set of thoughts and feelings, and a much greater range of knowledge and power, than it actually possesses. . . . It is therefore small wonder that his attitude towards the animal creation is one of reverence rather than superiority.

Only a lower man, one close to animals, would value them. Human rationalizations of this gap are analyzed in *A View to a Death in the Morning*, an elegant book on hunting by Matt Cartmill:

In policing the animal-human boundary, scientists have shown considerable ingenuity in redefining supposedly unique human traits to keep them from being claimed for other animals. Consider our supposedly big brains. Human beings are supposed to be smarter than other animals, and therefore *we* ought to have larger brains. But in fact, elephants, whales, and dolphins have bigger brains than ours; and small rodents and monkeys have relatively bigger brains (their brains make up a larger percentage of body weight than ours do). Scientists who study these things have accordingly labored to redefine brain size, dividing brain weight by basal metabolic rate or some other exponential function of body weight to furnish a standard by which these animals' brains can thus be deemed smaller than ours. The unique bigness of the human brain thus turns out to be a matter of definition.

This is not the only example of the manipulation of science toward the goal of dominance. In *The Mismeasure of Man*, Stephen Jay Gould cogently described the conscious or unconscious manipulation of data on brain size to prove that the scientist's racial group was inherently smarter than other groups. (A similar example of an attempt to force science into the service of racial discrimination may be found in Murray and Herrnstein's recent *The Bell*

Curve. This ugly piece of advocacy is depressing evidence that measurable intelligence is no guarantee of intelligent ideas.)

The Insensate Other

Animals' presumed lack of feeling has provided a major excuse for treating them badly. This has been so extreme that animals were long regarded as unable to feel pain, physical or emotional. But when an animal is hurt in a way that would hurt a person, it generally reacts much as a person would. It cries out, it gets away, then examines or favors the affected part, and withdraws and rests. Veterinarians do not doubt that wounded animals feel pain, and use analgesics and anesthetics in their practice. The only criterion that an animal fails to meet for feeling physical pain as humans understand it is the ability to express it in words. Yet the fish on the hook is said not to be thrashing in pain (or fear) but in a reflex action. A lobster in boiling water or puppies whose tails are being docked are said to feel nothing. A recent German book on animal consciousness argues to the contrary: "The fact that we so immediately understand these signals is just a further sign that we share with other animals the grand construction of our pain apparatus." When the subject is actually researched, the findings are in line with common sense: the apparent pain of the fish twisting on the hook is real.

It has always been comforting to the dominant group to assume that those in subservient positions do not suffer or feel pain as keenly, or at all, so they can be abused or exploited without guilt and with impunity. The history of prejudice is notable for assertions that lower classes and other races are relatively insensitive. Similarly, until the 1980s, it was routine for surgery on human infants to be performed with paralytic agents but without anesthesia, in the long-held belief that babies are incapable of feeling pain. It was believed, without evidence, that their nervous systems were immature. The notion that babies do not feel pain is directly counter to their screams and can only be classified as scientific myth. Yet it has been a tenet of human medicine, only recently acknowledged to be false in the wake of studies showing that infants who do not get pain medication take longer to recover from surgery.

A similar bigotry has extended to the presence of emotions in the poor, the foreign, those raised in impoverished or unenlightened cultures, and in children, who supposedly have not yet learned to feel in fully human ways. It is often asserted that when an infant smiles, for example, it is a physical response to gas in the intestines. The baby is said not to be smiling in response to other people, or out of happiness, but in response to digestive events. Despite the fact that adults do not smile as a result of discomfort in the stomach, this notion is widely repeated—though often not believed by the infant's parents. Studies showing that infant smiles are not correlated with burps, regurgitation, and flatulence have made little impact on this idea. Many people are gratified to think of infants as having diminished or no feelings.

If it is so easy to deny the emotional lives of other people, how much easier it is to deny the emotional lives of animals.

Anthropomorphism

The greatest obstacle in science to investigating the emotions of other animals has been an inordinate desire to avoid anthropomorphism. Anthropomorphism means the ascription of human characteristics—thought, feeling, consciousness, and motivation—to the nonhuman. When people claim that the elements are conspiring to ruin their picnic or that a tree is their friend, they are anthropomorphizing. Few believe that the weather is plotting against them, but anthropomorphic ideas about animals are held more widely. Outside scientific circles, it is common to speak of the thoughts and feelings of pets and of wild and captive animals. Yet many scientists regard even the notion that animals feel pain as the grossest sort of anthropomorphic error.

Cats and dogs are prime targets of anthropomorphism, both wrongly and rightly. Ascribing unlikely thoughts and feelings to pets is common: "She understands every word you say." "He sings his little heart out to show how grateful he is." Some people deck

reluctant pets in clothing, give them presents in which they have
no interest, or assign their own opinions to the animals. Some dogs
are even taught to attack people of races different from their owners'. Many dog lovers seem to enjoy believing that cats are selfish,
unfeeling creatures who heartlessly use their deluded owners, compared with loving, loyal, and naive dogs. More often, however,
people have quite realistic views about their pets' abilities and attributes. The experience of living with an animal often provides a
strong sense of its abilities and limitations—although even here, as
for people living intimately with people, preconceptions can be
more persuasive than lived evidence, and can create their own reality.

Consider three statements about a dog's behavior: "Brandy's
upset because we forgot her birthday," "Brandy feels left out and
wants your attention," and "Brandy is performing the submissive
display of a low-ranking canid." The first two statements can both
be called anthropomorphic and the last is the jargon of ethology,
the scientific study of animal behavior. The first statement is probably anthropomorphic error or projection; the speaker would feel
bad if his or her birthday were forgotten and assumes the dog feels
the same, but of course the idea that the dog knows what birthdays
and birthday parties are is far-fetched. The third statement describes an "ethogram" of the dog's actions and avoids any mention
of thought or feeling. It is an incomplete description, one that
deliberately describes events and avoids explaining them, restrict-
ing its own predictive ability. The second statement interprets the
dog's feelings. While it could be mistaken, it is only anthropomorphic if dogs cannot feel left out and cannot want attention—which
most dog owners know to be untrue. In the end, it may be the most
useful of the three statements.

Perhaps the richest source of anthropomorphic error occurs in
human thinking about wild animals. Since people live with domestic animals, erroneous theories about their behavior are likely to be
disproved in the course of events. But since most people's contact
with wild animals is so limited, theories about them may never run
up against facts, and we remain free to imagine ravening wolves,
saintly dolphins, or crows who follow parliamentary procedure.

Science considers anthropomorphism toward animals a grave mistake, even a sin. It is common in science to speak of "committing" anthropomorphism. The term originally was religious, referring to the assigning of human form or characteristics to God—the hierarchical error of acting as though the merely human could be divine—hence the connotation of sin. In the long article on anthropomorphism in the 1908 *Encyclopedia of Religion and Ethics,* the author (Frank B. Jevons) writes: "The tendency to personify objects—whether objects of sense or objects of thought—which is found in animals and children as well as in savages, is the origin of anthropomorphism." Men, the idea goes, create gods in their own image. The best-known example of this tendency comes from the Greek author Xenophanes (fifth century B.C.). He notes that Ethiopians represent the gods as black, Thracians depict them as blue-eyed and red-haired, and "if oxen and horses . . . had hands and could paint," their images of gods would depict oxen and horses. The philosopher Ludwig Feuerbach concluded that God is nothing but our projection, on a celestial screen, of the essence of man. In science, the sin against hierarchy is to assign human characteristics to animals. Just as humans could not be like God, now animals cannot be like humans (note who has taken God's place).

Anthropomorphism as Contagion

Young scientists are indoctrinated with the gravity of this error. As animal behaviorist David McFarland explains, "They often have to be specially trained to resist the temptation to interpret the behavior of other species in terms of their normal behavior-recognition mechanisms." In his recent book *The New Anthropomorphism,* behaviorist John S. Kennedy laments, "The scientific study of animal behavior was inevitably marked from birth by its anthropomorphic parentage and to a significant extent it still is. It has had to struggle to free itself from this incubus and the struggle is not over. Anthropomorphism remains much more of a problem than most of today's neobehaviorists believed. . . . If the study of animal behavior is to mature as a science, the process of liberation from the delusions of anthropomorphism must go on." His hope is that "anthropomorphism will be brought under control, even if it cannot be cured completely. Although it is probably programmed into us genetically as well as being inoculated culturally that does not mean the disease is untreatable."

The philosopher John Andrew Fisher has noted, "The use of the term 'anthropomorphism' by scientists and philosophers is often so casual as to almost suggest that it is a term of ideological abuse, rather like political or religious terms ('communist' or 'counterrevolutionary') that need no explication or defence when used in criticism."

In a science dominated by men, women have been deemed especially prone to empathy, hence anthropomorphic error and contamination. Long considered inferior to men precisely on the ground that they feel too much, women were thought to overidentify with the animals they studied. This is one reason why male scientists for so long did not encourage female field biologists. They were too emotional; they allowed emotions to sway judgments and observations. Women, it was felt, were more likely than men to attribute emotional attitudes to animals by projecting their own feelings onto them, thereby polluting data. Thus did gender bias and species bias converge in a supposedly objective environment.

To accuse a scientist of anthropomorphism is to make a severe criticism of unreliability. It is regarded as a species-confusion, a forgetting of the line between subject and object. To assign thoughts or feelings to a creature known incapable of them would, indeed, be a problem. But to ascribe to an animal emotions such as joy or sorrow is only anthropomorphic error if one knows that animals cannot feel such emotions. Many scientists have made this decision, but not on the basis of evidence. The situation is not so much that emotion is denied but that it is regarded as too dangerous to be part of the scientific colloquy—such a minefield of subjectivity that no investigation of it should take place. As a result, any but the very most prominent scientists risk their reputations and credibility in venturing into this area. Thus many scientists may actually believe that animals have emotions, but be unwilling not only to say that they believe it, but unwilling to study it or encourage their students to investigate it. They may also attack other scientists who try to use the language of the emotions. Nonscientists who seek to retain scientific credibility must tread carefully. An administrator at one internationally known animal training institute remarked, "We don't take a position on whether animals have emotions, but I'm sure if you talked to *any* one of us we'd say, 'Sure they have emotions.' But as an organization we would not want to be depicted as saying they have emotions."

Linguistic Taboos

From the belief that anthropomorphism is a desperate error, a sin or a disease, flow further research taboos, including rules that dictate use of language. A monkey cannot be angry; it exhibits aggression. A crane does not feel affection; it displays courtship or parental behavior. A cheetah is not frightened by a lion; it shows flight behavior. In keeping with this, de Waal's use of the word *reconciliation* in reference to chimpanzees who come together after a fight has been criticized: Wouldn't it be more objective to say "first postconflict contact"? In the struggle to be objective, this kind of language employs distance and the refusal to identify with another creature's pain.

Against this scientific orthodoxy, the biologist Julian Huxley has argued that to imagine oneself into the life of another animal is both scientifically justifiable and productive of knowledge. Huxley introduced one of the most extraordinary accounts of a deep and emotional tie between a human being and a free-living lioness, Joy Adamson's *Living Free*, as follows:

> When people like Mrs. Adamson (or Darwin for that matter) interpret an animal's gestures or postures with the aid of psychological terms—anger or curiosity, affection or jealousy—the strict Behaviourist accuses them of anthropomorphism, of seeing a human mind at work within the animal's skin. This is not necessarily so. The true ethologist must be evolution-

minded. After all, he is a mammal. To give the fullest possible interpretation of behaviour he must have recourse to a language that will apply to his fellow-mammals as well as to his fellow-man. And such a language must employ subjective as well as objective terminology—*fear* as well as *impulse to flee*, *curiosity* as well as *exploratory urge*, *maternal solicitude* in all its modulations in welcome addition to goodness knows what complication of behaviourist terminology.

Huxley's argument ran counter to mainstream scientific thinking when he wrote that in 1961, and it remains so today. A contemporary example is provided by Alex, the African grey parrot, who was being trained or tested by experimenters who varied the requests they made of him to avoid cueing and to prevent Alex from becoming bored. When reviewers of a paper that researcher Irene Pepperberg submitted to a scientific journal vetoed her use of the term *boredom*, she remarked:

> I had a referee go ballistic on me. And yet, you've watched the bird, he looks at you, he says "I'm gonna go away." And he walks! The referee said that was an anthropomorphic term that had no business being in a scientific journal. . . . I can talk in as many stimulus-response type terms as you want. It turns out, though, that a lot of his behaviors are very difficult to describe in ways that are not anthropomorphic.

What is wrong with exploring the idea, based on many such observations in a research setting, that parrots and humans may have a shared capacity for boredom?

Naming

In the study of animal behavior it has long been taboo for scientists to name the animals. To separate individuals, they might be called Adult Male 36, or Juvenile Green. Most field workers over the generations have resisted this precept, naming the animals they spent their days watching, at least for their own use, Spot-Nose and Splotch-Tail, Flo and Figan, or Cleo, Freddy, and Mia. In their published work, some reverted to more remote forms of identification; others continued to use names. Sy Montgomery reports that, in 1981, anthropologist Colin Turnbull declined to provide a supporting statement for Dian Fossey's book of observations on mountain gorillas because she assigned names to the gorillas. It is even more common not to name animals in laboratories, perhaps for the same reason that farmers often avoid naming animals they expect to slaughter: proper names have a humanizing effect, and it is harder to kill a friend.

Rebutting the view that naming animals only causes one to assign them human traits, elephant researcher Cynthia Moss notes that the opposite happens to her: people remind her of elephants. "When I am introduced to a person named Amy or Amelia or Alison, across my mind's eye flashes the head and ears of that elephant." The no-name norm has gradually changed, particularly among primatologists, perhaps because of the outstanding work of researchers who named—and admitted that they named—the subjects they studied. Bekoff and Jamieson, a field biologist and a philosopher, have argued that it is not only permissible but advisable to name animals under study, since empathy increases understanding. Yet as recently as 1987, researchers studying elephants in Namibia (then South-West Africa) were instructed by park authorities to assign the animals numbers because names were too sentimental. Granted that a number is more dehumanized than a name, does that make it more scientific? Assigning names to them—referring to a chimpanzee as Flo or Figan—can be called anthropomorphic, but so is assigning numbers. Chimpanzees are no more likely to think of themselves as F2 or JF3 than as Flo or Figan.

We do not know if animals name themselves or each other. We do know that animals recognize other animals as individuals and distinguish between them. Names are the way humans label such distinctions. Bottle-nosed dolphins may identify and imitate one another's signature whistles, something very close to a name. A similar phenomenon has been observed in captive birds. When their mate was removed, caged ravens and Shama thrushes "frequently uttered sounds or song elements which were otherwise principally or exclusively produced by the partner. On hearing these sounds, the bird so 'named' returned at once, whenever this was possible." The ability to call a mate by name could be even more useful to wild birds. Some animals clearly respond emotionally to being given a name. Mike Tomkies in *Last Wild Years* writes that "only the ignorant pour scorn on this habit of mine of giving names to the creatures that, over the years, have shared my home. And also others that have not. So long as it is not a harsh sound, it matters little what the name is, but there can be no doubt whatever that an animal or bird will respond differently, become more trusting, once it is given a name."

If naming the animals one studies promotes empathy toward them, this may help rather than occlude insight into their natures. The essential fact glossed over in the attack on anthropomorphism is that humans are animals. Our relation to animals is not a literary exercise in creating charming metaphors. As the philosopher Mary Midgley puts it: "The fact that some people are silly about animals cannot stop the topic being a serious one. Animals are not just one of the things with which people amuse themselves, like chewing-gum and water-skis, *they are the group to which people belong*. We are not just rather like animals; we *are* animals." To act as if humans are a completely different order of beings from other animals ignores the fundamental reality.

Anthropomorphism Without Really Meaning It

Even fierce opponents of anthropomorphism concede that it often works in trying to predict animal behavior. By considering what an animal feels or thinks, we may improve our ability to project how it will act. Such guesses have a high success rate. While successful prediction does not prove that the animal actually felt or thought what was imagined, it is a standard test of scientific theories. John S. Kennedy, the animal behaviorist who views anthropomorphism as a disease, concedes nevertheless that it is a useful way to predict behavior. Kennedy argues that anthropomorphism works because animals have evolved to act *as if* they thought and felt: "it is natural selection and not the animal that ensures that what it does mostly 'makes sense,' as we are wont to say."

Even though Kennedy disavows the "assumptions that they have feelings and intentions," he acknowledges that empathy can be useful for generating questions and making predictions. Thus one might predict that a cheetah, fearful for her cubs, may run close to a lion to lure it away. Under Kennedy's formulation, if the cheetah does so, it does not mean she fears for the lives of her cubs. It only means that she has evolved to act *as if* she fears for their lives. To speculate that leaving more offspring is the ultimate cause of her behavior is permitted. Not permitted is to speculate that fear for their lives is its proximate cause, far less about how she may feel seeing the lion grabbing them. Why is it so impossible to know what animals feel, no matter how much or what kind of evidence there is? How is knowing about their feelings different, in truth, from the assumptions made routinely about the feelings of other people?

The Solipsistic Defense

Short of being another person, there is no way to know with certainty what another person feels, although few people, even philosophers, carry their solipsism (the belief that the self can know nothing but the self) this far. In learning others' feelings, people are not always led by words alone, but watch behavior—gestures, the face, the eyes—patterns and consistency over time. Conclusions are based on this, and ground everyday life decisions. We love certain people, hate others, trust some, fear others, and act on this basis. Belief in the emotions of others is indispensable to life in human society. N. K. Humphrey writes, "For all I know no man other than myself has ever experienced a feeling corresponding to my feeling of hunger; the fact remains that the concept of hunger, derived from my own experience, helps me to understand other men's eating behavior." On human claims not to know animals' pain, Midgley has said of the extreme solipsistic position: "If a torturer excused her activities by claiming ignorance of pain on the grounds that nobody knows anything about the subjective sensation of others, she would not convince any human audience. An audience of scientists need not aim at providing an exception to this rule." She locates the basis of human assumptions of natural superiority underlying the position of the solipsist when she quotes an astonishing passage from *Ethics*, by the seventeenth-century Dutch philosopher Benedict de Spinoza:

> It is plain that the law against the slaughtering of animals is founded rather on vain superstition and womanish pity than on sound reason. The rational quest of what is useful to us further teaches us the necessity of associating ourselves with our fellow-men, but not with beasts, or things, whose nature is different from our own; we have the same rights in respect to them as they have in respect to us. Nay, as everyone's right is defined by his virtue, or power, men have far greater rights over beasts than beasts have over men. Still I do not deny that beasts feel; what I deny is that we may not consult our own advantage and use them as we please, treating them in the way which best suits us; for their nature is not like ours, and their emotions are naturally different from human emotions.

Spinoza refrains from discussing how he knows that animal emotions are different from human ones, or from explaining how this justifies the human exploitation, plunder, and murder of them. He simply says we have more power than they do. Might makes right. José Ortega y Gasset's defense of hunting comes to the same conclusion, insisting that the victim is always asking for it:

> [Hunting] is a relationship that certain animals impose on man, to the point where not trying to hunt them demands the intervention of our deliberate will. . . . Before any particular hunter pursues them they feel themselves to be possible prey, and they model their whole existence in terms of this condition. Thus they automatically convert any normal man who comes upon them into a hunter. *The only adequate response to a*

being that lives obsessed with avoiding capture is to try to catch it.
[Ortega y Gasset's italics]

Such delusional anthropomorphism, based in turn on a human model that itself is delusional, reveals deep and hidden assumptions and interests. Ortega y Gasset's buried premise—that hunted beings seek their own demise—closely resembles rationales about rape. A common excuse of rapists is that women ask for rape, thus seeking and causing their own violation, most especially when actively trying to avoid it. A similar exoneration of hunters is sought here by justifying the capture of animals by calling animal flight from capture an "obsession"—meaning they most desire what they most strenuously flee.

Simpler forms of anthropomorphism can also interfere with observation and distort understanding. Carolus Linnaeus, the eighteenth-century Swedish naturalist who developed the classification system of living things, wrote of the frog: "These foul and loathsome animals are . . . abhorrent because of their cold bodies, pale color, cartilaginous skeletons, filthy skin, fierce aspect, calculating eye, offensive smell, harsh voice, squalid habitation and terrible venom." The words are all emotive, referring to emotions Linnaeus felt when *he* saw a frog. They are pure projection. *Calculating* is not a scientific term to describe a frog's eye. This passage is art—it describes little in the physical world, but powerfully conveys the scientist's subjective state.

Assigning Human Gender Roles to Animals

Another problem with anthropomorphism has been that human views of gender—often as wrong as human views of animals—have been attributed to animals. People sometimes expect a male animal to lead the herd or be dominant or more aggressive even in species where the reality is different. A recent nature program on television featured a family of cheetahs in Tanzania's Serengeti National Park. The male cub was called Tabu and the female Tamu—Swahili for Trouble and Sweetness. One expects different things from a Sweetness than a Trouble. Surely the sentence "Trouble is prowling around my tent" is more threatening than "Sweetness is prowling around my tent." Sociobiology has tended to encourage prejudices men have about women by insisting that they are "natural," by which is meant they can be found among members of the animal kingdom. As already noted, one can prove almost anything by careful choice of species. It does not seem accidental that human society has for so long been compared to baboon society, despite the facts that baboons are far more sexually dimorphic than humans and that baboons do not form mated pairs. The idea seems to be to impose greater gender inequality on human females by enforcing a supposedly natural template.

A serious problem with careless human-animal comparisons is the inadequacy of our present knowledge of animals' lives, especially of crucial matters like the role of culture in animal learning in the wild. Elephants, for example, learn from their elders which humans to fear based on the history of the herd with humans. Mike Tomkies describes watching an eaglet in the wild being taught to fly so as to hunt and kill by repeated demonstrations on the part of its parent, who was clearly showing the youngster what to do rather than engaging itself in search of prey. Evidently the eaglet is not born knowing this. It is transmitted by learning, that is, by culture. It is natural, but it is also learned; that it has to be learned does not make it unnatural. To use the word *natural* to describe how the eaglet kills simply means that an animal was observed doing it. The distinction between innate and natural, on the one hand, and cultural and learned, on the other, loses much of its force in light of more recent observations on what animals teach each other.

Anthropocentrism

The real problem underlying many of the criticisms of anthropomorphism is actually *anthropocentrism*. Placing humans at the center of all interpretation, observation, and concern, and dominant men at the center of that, has led to some of the worst errors in science, whether in astronomy, psychology, or animal behavior. Anthropocentrism treats animals as inferior forms of people and denies what they really are. It reflects a passionate wish to differentiate ourselves from animals, to make animals other, presumably in order to maintain humans at the top of the evolutionary hierarchy and the food chain. The notion that animals are wholly other from humans, despite our common ancestry, is more irrational than the notion that they are like us.

But even if they were not like us at all, that is no reason to avoid studying them for their own sakes. The point has been made by J. E. R. Staddon that "psychology as a basic science should be about intelligent and adaptive behavior, wherever it is to be found, so that animals can be studied in their own right, for what they can teach us about the nature and evolution of intelligence, and not as surrogate people or tools for the solution of human problems." The knowledge obtained from such study, whether or not it contributes to the solution of human problems, is still knowledge.

Animals as Saints and Heroes

Idealizing animals is another kind of anthropocentrism, although not nearly as frequent as their denigration and demonization. The belief that animals have all the virtues to which humans aspire and none of our faults is anthropocentric, because at its core is an obsession with the repulsive and wicked ways of humans, which animals are used to highlight. In this sentimental formulation the natural world is a place without war, murder, rape, and addiction, and animals never lie, cheat, or steal. This view is embarrassed by reality. Deception has been observed in animals from elephants to arctic foxes. Ants take slaves. Chimpanzees may attack other bands of chimpanzees, unprovoked and with deadly intent. Groups of dwarf mongooses battle other groups for territory. The case of the chimpanzee murderers Pom and Passion, who killed and ate the infants of other chimpanzees in their group, has been well documented by Jane Goodall's research team. Orangutans have been seen to rape other orangutans. Male lions, when they join a pride, often kill young cubs who were fathered by other lions. Young hyenas, foxes, and owls have been seen to kill and eat their siblings.

All is not as humans wish it to be among our evolutionary cousins. One has to sympathize with Jane Goodall's reaction to some chimpanzees' treatment of one old animal, his legs wholly paralyzed by polio, who was lonely, shunned, and sometimes attacked by those who were still healthy. In the hope of inducing companions who were grooming each other to groom him as well, he dragged himself up into a tree:

> With a loud grunt of pleasure he reached a hand towards them in greeting—but even before he made contact they both swung quickly away and, without a backward glance, started grooming on the far side of the tree. For a full two minutes, old Gregor sat motionless, staring after them. And then he laboriously lowered himself to the ground. As I watched him sitting there alone, my vision blurred, and when I looked up at the groomers in the tree I came nearer to hating a chimpanzee than I have ever done before or since.

It is hard to romanticize anything this ugly.

It has been a long time since anyone called the lion the king of beasts (except in a Walt Disney film), but dolphins have recently been romanticized as smarter, kinder, nobler, more pacific, and better at living in groups than people. This ignores the well-documented fact that dolphins can be quite aggressive. Recently it has been discovered that some dolphins occasionally rape. At the same time, animal cruelty does not approximate the human standard. It is unlikely that dolphin rape rivals the human figures. One respected random sample study in 1977 found that almost half of all the women in one U.S. city had been victims of rape or attempted rape at least once in their lives. Child abuse may occur rarely in the wild, but nothing compares with over one in every three girls being sexually abused as children, as shown in a major American study conducted in 1983 by the same researcher.

Zoomorphism

If humans can misunderstand animals by assuming they are more like us than they are, can animals also wrongly project their feelings onto us? Do animals commit what might be called zoomorphism, ascribing their attributes to humans? A cat who brings a human offerings of dead rodents, lizards, and birds day after day, no matter how often these objects are greeted with loathing, commits zoomorphism. This is the equivalent of offering candy to a cat, as children sometimes do. In *The Hidden Life of Dogs*, Elizabeth Marshall Thomas writes: "When a dog with a bone menaces a human observer, the dog actually assumes that the person wants the slimy, dirt-laden object, and is applying dog values, or cynomorphizing." Were a dog to give a history of the human race, some valuable attributes may be denied us, just as our history of any animal civilization would doubtless miss many of its signal achievements.

Fear, Hope, and the Terrors of Dreams

Animal behaviorists are unlikely to acknowledge that terror can return in the dreams of animals. And yet from a Kenyan "elephant orphanage" comes a report of baby African elephants who have seen their families killed by poachers, and witnessed the tusks being cut off the bodies. These young animals wake up screaming in the night. What else but the nightmare memories of a deep trauma could occasion these night terrors?

Wildlife biologist Lynn Rogers has spent decades studying black bears, following them through forests and swamps. As a graduate student he learned about black bears from his professor, Albert Erikson. One day they were trying to take a blood sample from an anesthetized wild bear, when it suddenly woke. The bear lunged at Erikson. To Rogers's surprise, Erikson lunged back. The bear turned to Rogers. Erikson said, "Lunge!" Rogers obediently lunged at the bear, who turned and ran away. Rogers says, "I was learning things that would help me interpret bears' actions in terms of their own fear rather than mine."

An error to which anthropomorphism can lead is to see bears through our own emotions: we fear them, so we perceive them as angry and hostile. The equal and opposite error into which the fear of anthropomorphism can lead is to refuse to recognize that bears can feel their own emotions. Rogers learned to observe bears in terms of those emotions, discovering that the bears themselves were often fearful. He learned what frightened them and how not to frighten them. "Once I started looking at bears in terms of their fear, and interpreted all the things that used to scare me and interpreted those in terms of the bear's fear, it was easy to gain their trust and begin walking with them very closely, sleeping with them —doing all the things that you have to do to see how an animal really lives in its world." So well did Rogers learn to understand the wild bears that he could curl up for the night a few feet from their den or even handle their cubs. Asked whether scientists don't usually avoid using words like *fear* and *trust* to describe animal behavior, he replied, "Yes. But I think that we miss the mark more by ignoring those emotions than by taking them into account. Those are basic emotions that animals and people share."

His description of a suddenly alarmed bear shows how humans can learn to "read" bears: "You can be very close to a bear and have things be calm, until some little unidentified noise happens far off in the forest. Then the bear is suddenly keyed up, wary. . . . Whenever there's anything that makes the bear take a deep breath, which is the first sign of their fear, and then you see its ears prick up, you think, 'Better give the bear a little bit more room, don't be standing right on top of it, because there's a good chance it'll whack you,' " says Rogers cheerfully. "It feels threatened by some other thing and it wants room and the peace of mind from you to deal with that. After being told in no uncertain terms by bears to get away in that situation, after a while I learned."

A Cornerstone Emotion

Of all emotions animals might feel, fear is the one that skeptics most often accept and one of the few that comparative psychology investigates. One reason is that fearfulness has an obvious evolutionary advantage. Fear can serve as a mechanism to trigger defensive behavior, so its survival value is clear for any organism capable of defense. Fear can set animals running, diving, hiding, screaming for help, slamming their shells shut, bristling their quills, or baring their teeth. If an animal had no mode of defense, fear would confer no benefit. Yet fear has also been known to interfere with survival: the actions of a panicking person or animal are not always the wisest, as when a terrified soldier on a battlefield runs into the line of fire.

People also find it easy to believe that animals feel fear because this emotion is one that humans often elicit from animals, and may even enjoy eliciting. An urban dweller who has never so much as visited a zoo has probably scattered birds into flight, shooed insects away, seen cats flee from dogs or dogs flee from bigger dogs, and has no reason to doubt that animals feel fear.

Nor does a powerful intellect seem necessary to experience fright. Intellect may help one detect subtler reasons to fear, but the less intelligent still find plenty to fear. Those who wish to believe that a great gulf separates people from other animals seldom seem threatened by the notion of animal fear. In animals it may not be called an emotion, however. Thus, while dictionaries call fear an emotion, animal behaviorists may prefer the definition of fear that appears in *The Oxford Companion to Animal Behavior:* "a state of motivation which is aroused by certain specific stimuli and normally gives rise to defensive behavior or escape."

The Picture of Terror

The biological traces of fear are easy to find in a laboratory. (Indeed, what animal would not have reason to fear a laboratory?) A small electrical impulse to a cat's amygdala (part of the brain's limbic system) produces alertness, a larger one produces the expressions and actions of terror. A rat whose amygdala has been removed loses the fear of cats and will walk right up to one. Researchers at New York University trained rats to expect an electric shock when they heard a tone, and discovered to their surprise that the nerve impulses in the rats taught to fear the tone went straight from the ear to the amygdala, instead of via the usual route through the auditory cortex. The theory is that the amygdala attaches emotional import to some forms of learning. Endocrine studies show that hormones such as epinephrine and norepinephrine help pass along fear messages. Geneticists say that in just ten generations of breeding, two strains of rats can be produced from a parental stock, one fearful, one calm.

But even biologists concede that physiological symptoms alone do not form a complete description of fear. Philosopher Anthony Kenny has given the example of a person who fear heights and avoids them scrupulously, as compared to a relatively fearless mountain climber. The person who avoids heights may succeed in doing so and, as a result, seldom exhibit physiological signs of fear. The climber, more often at risk, may show such signs more often, yet cannot be said to be more afraid of heights. Perhaps, though, the notion of a "counterphobia," developed by the psychoanalyst Otto Fenichel, is not entirely inappropriate here. He spoke of people seeking out the very thing they most fear because the fear is unconscious. Thus at least some climbers are terrified of heights but cannot acknowledge this fear to themselves. Their behavior is a kind of deep overcompensation, an internal self-deception meant to keep the feared but fascinating object in constant view. Is this, like the well-known compulsion to repeat traumas, a search for mastery?

Perhaps counterphobia is not confined to humans. Many animals of species that are frequently preyed upon show a macabre interest in the deaths of others like them. Studying hyenas in the Serengeti, Hans Kruuk was struck by the frequency with which hyenas or other predators at a kill were closely observed by wildebeest or gazelle who had drifted over to watch. This has been called "behavior of fascination" or "the bystander phenomenon." Such onlookers are attracted even when the victim is not of their own species. Animals of prey species also show interest in predators who are not at kills, watching and even following them. A cheetah being observed by a crowd of gazelle made a sudden dart and caught one of them, so the behavior has risks. Kruuk speculates that this dangerous behavior conveys a selective advantage—

either because it is worthwhile for prey species to keep an eye on the predator, preventing ambushes, or because they learn valuable information about predators. In his classic study of red deer, F. Fraser Darling noted that "deer have a marked objection to allowing any person or object out of their sight which they may think to be a source of danger." It may also be an example of counterphobia.

In *The Expression of the Emotions in Man and Animals,* Darwin made a systematic study of how animals look when they are afraid. In both humans and animals, he found, some or all of the following may occur: the eyes and mouth open, the eyes roll, the heart beats rapidly, hairs stand on end, muscles tremble, teeth chatter, and the sphincter loosens. The frightened creature may freeze in its place or cower. These rules hold true across a remarkable array of species. Somehow it is surprising to learn that when dolphins are terrified, their teeth chatter and the whites of their eyes show, or that a frightened gorilla's legs shake. Such familiar behavior in a wild animal is a reminder of our ultimate kinship. Melvin Konner has written, "We are—not metaphorically, but precisely, biologically—like the doe nibbling moist grass in the predawn misty light; chewing, nuzzling a dewy fawn, breathing the foggy air, feeling so much at peace; and suddenly, for no reason, looking about wildly."

Other symptoms of fear may be more specific to a species. A frightened mountain goat, biologist Douglas Chadwick reports, flattens its ears, flicks its tongue past its lips, crouches and raises its tail. According to Chadwick, a kid raises its tail when it wants attention or to nurse. The adult continues to raise its tail when fearful. If the tail is partially raised, Chadwick says, it means "I'm worried," and a completely erect tail means "I'm scared," or maybe "Help, Momma!"

Aviculturalist Wolfgang de Grahl notes that frightened young grey parrots in new surroundings may not only flutter wildly at the approach of humans but may hide their heads in a far corner. In de Grahl's view, these birds probably believe, like the ostriches who were once said to hide their heads in the sand, that they cannot be seen when they do so. But this is likely based upon an overestimation of the stupidity of birds. Humans who cover their eyes or turn their faces away from scary sights do not believe they cannot be seen. Perhaps, like humans, the parrots cannot bear the sight of what frightens them, or are trying to keep their feelings from overwhelming them.

What Animals Fear

Because people have lived and worked with horses for so long, some of the things that frighten them are fairly well understood. In addition to such obvious dangers as predators, they may be alarmed by unfamiliar motions, noises, and smells. Changes in their environment often frighten horses. Skittish horses even appear to be alarmed by imagined changes: an object that a horse has passed countless times may suddenly cause a horse to shy although there has been no alteration. What frightens one horse will leave another horse unmoved, and some horses seldom display fear. Horses may also be afraid to go to places that do not smell as if horses have been there. A horse that goes happily into horse trailers will sometimes refuse to go into a brand-new horse trailer.

Personal history also plays a part in the genesis of fear for a particular animal, who can learn to fear something that it did not fear before. This is expressed in commonsense beliefs. For example, if you pick up a stick to toss for a dog to retrieve and instead it cringes in fear, your first thought is likely to be that the dog has been beaten. Animals form associations of fear with objects that have frightened them in the past. Memories can be triggered by resemblances or perhaps even by wandering thoughts.

Animals also learn fear to avoid pain. Laboratory rats fear pain and learn to fear receiving an electric shock. Coyotes learn to fear getting a faceful of porcupine quills. Monkeys learn that a long fall is painful.

Fear and Self-Defense

Most animals fear their predators, logically enough. How they recognize them as predators if they have never seen them in action is not always clear, but the reaction is. In the Rockies, Chadwick one day saw a lynx stalking a large mountain goat billy. It sneaked to a ledge above the billy, to a position that seemed perfect for pouncing, but hesitated. Then the goat spotted the lynx, and backed off into a corner. After a while the goat came forward, stamped his feet, and began leaping up in the direction of the cat, hooking toward it with his horns. The lynx watched for a while, now and then dangling a paw toward the goat before finally walking off. It appeared that the goat was frightened of the predatory lynx at first, but then lost its fear and became aggressive. The lynx was mildly frightened of the goat—enough not to attack immediately and eventually to give up.

One factor in the recognition of predators may be an innate response to staring eyes. Birds have been found to be more likely to mob a stuffed owl if it has eyes. Young chicks who have never seen a predator avoid objects with eyes or eye-spots on them, particularly if the eyes are large. Wild birds at a feeder table are much more apt to flee if a design that is highlighted on the feeder resembles eyes, and the more realistic the eyes, the greater their panic.

Fear of falling from heights also appears to be innate in many animals. Infants of many species (including humans) show terror when confronted with a steep drop, or the convincing image of a steep drop, even if they have never been dropped or seen a drop. The fear of heights is probably triggered in some species more easily than in others. A creature that lives in high places cannot survive if it spends too much time quaking with alarm. Yet the mountain goats observed by Chadwick also showed signs of fear when searching for a foothold on cliffs too precipitous even for them, or when loose rock underfoot started to slide toward a drop.

A brown bear cub who fell into the McNeil River in Alaska and was carried out into the rapids showed signs of fear—flattened ears and wide, rolling eyes. His mother saw him fall in but did not seem alarmed, only going after him once he had been carried a considerable distance away. Perhaps she did not realize that what was safe for her was not safe for him. Or perhaps she realized, or felt, that he was in no real danger. The cub succeeded in getting out on his own.

Alone and Lost

For social animals, or for the young of most species, loneliness holds fears. The fear of being alone is sometimes hard to separate from the fear of being lost. Wingnut, a particularly timid brown bear cub observed on the McNeil River, was said by Thomas Bledsoe to be literally afraid of his own shadow. He was afraid of being left alone, and would call "hysterically" every time his mother left to fish, continuing until she returned. Again, it may well be that an earlier experience lies behind this reaction. A Pacific bottle-nosed porpoise, Keiki, who lived in a marine park, was released into a bay nearby. Separated from his companions into a location he did not know, Keiki was stricken with terror, teeth chattering and eyes rolling.

Zoo keepers report that captive elephants are subject to "sudden-death syndrome" or "broken-heart syndrome," which happens (most often with young elephants) when they are separated from their social group or put in a new enclosure by themselves. Jack Adams of the Center for the Study of Elephants ascribes this to "gripping fear."

Like the horses afraid of unusual things, captive but untamed grey parrots are suspicious of changes in their surroundings. Rather than eat from a new bowl, they will go hungry for days. Even once they have learned to trust and accept food from a particular individual, a change of clothing can create alarm. One aviculturalist reported that a group of mistrustful parrots would accept peanuts only from the aviculturalist's mother—and only when she wore her usual apron. The term *neophobia* has been coined for this fear of the unfamiliar. Neophobia can produce odd reactions in an animal raised in unusual circumstances. Indian conservationist Billy Arjan Singh raised an orphaned leopard cub and a tiger cub.

In each case the young cat was terrified by its first glimpses of the jungle and had to be patiently soothed, repeatedly taken for walks in the jungle, and generally convinced that it was worth visiting.

Cody, an orangutan raised from infancy by humans, was stricken with terror on first beholding another orangutan. The hair stood up all over his body. He recoiled in fear and hid behind his human "parent," clinging so hard that he left marks. The placid orangutan who frightened him so much happened to be his own mother.

Jim Crumley has described watching a flock of two hundred whooper swans resting in a field in Scotland. As he looked on, a wave of disturbance passed through the flock. Sleeping birds raised their heads and stood looking to the west, but then the flock settled down. The swans gradually relaxed, and then suddenly became agitated again: all heads shot up, and they called to each other in alarm. This happened three times before the perplexed Crumley understood what was worrying the swans. A thunderstorm was approaching, and the swans had heard it coming before he did. He watched them through the ensuing storm, and saw that the flashes of lightning brought no reaction, but that every clap of thunder terrified them.

Learning to Be Afraid

Much fear is learned. This accords with classical behavioral conditioning theory, in which animals, including humans, learn to associate negative stimuli with particular events. It is important to be wary of too readily explaining things as innate or instinctive, when they might just as easily be learned by hard experience, or even somehow taught by other members of the species. Elizabeth Marshall Thomas notes some of the specific fears of a husky, Koki, she acquired as an adult animal. The sound made by an object whizzing through the air, such as a rope or a stick, would cause Koki to cower, with chattering teeth and hair on end. According to Marshall Thomas, "the sound of alcohol in a man's voice" had the same effect. It is possible that Koki was reacting to scent rather than sound, since alcohol affects the odor of human perspiration; but in any case, she had learned to be frightened of men who had been drinking. It is hard to avoid thinking she had been hit by a drunken man.

Classical conditioning theory was shaken up when it was discovered that some stimuli are far more easily associated with fear than others. Rats readily associate food with illness, and will avoid a food if they have been ill after eating it. But they are very unlikely to associate an electrical shock or a loud noise with illness, no matter how often experimenters pair the two stimuli. Many people are afraid of snakes or spiders who have never had a bad experience with them and who seldom see them. Yet, as Martin Seligman has pointed out, very few people have phobias about hammers or knives, although they are much more likely to have been injured by these. Perhaps their familiarity with other uses of these objects dulls fear.

Mountain goats have learned to fear avalanches or rock slides and to take evasive action. When goats hear the rumble of a slide overhead, they put their tails up and their ears back and run for a sheltering overhang, if one is nearby. If not, they stamp, crouch, and press themselves against the mountain. Some goats take off at the last moment.

Nameless Fears

Everyone has experienced fear without an apparent object—the sense that an unknown misfortune impends. At other times, the fear is in response to the sense that we are on unfamiliar ground, like Singh's tiger and leopard cubs. We feel that something bad could happen, though we don't know what. Fear can exist without an object, a vertigo of the morale.

In Hwange National Park in Zimbabwe the elephants are culled annually. During this culling, elephant family groups are herded by aircraft toward hunters who shoot all except the young calves, who are rounded up for sale. The elephant calves run around, scream, and search for their mothers. One year a wildlife

guide at a private sanctuary ninety miles away from the park noticed that eighty elephants vanished from their usual haunts on the
day culling started at Hwange. He found them several days later,
bunched at the end of the sanctuary as far from the park as they
were able to get.

It has been discovered quite recently that elephants can communicate over long distances by means of subsonic calls—sounds
pitched too low for people to hear. So it is not surprising that the
sanctuary elephants apparently received some frightening message
from the Hwange elephants. But unless elephant communication is
far more refined than anyone has yet speculated, the message cannot have been very specific. The sanctuary elephants must have
known that something very bad was happening to Hwange elephants, but they can hardly have known what it was. The object of
their fear was inchoate, but the fear was real.

Fear for Others

Humans not only fear for themselves, but may fear for others.
This feeling borders on empathy, something people are much less
likely to concede to animals than fear. While examples of animals
frightened for themselves are numerous, examples of fear for others are scarcer. Often the situation is equivocal: a monkey that
shows physical signs of fear when watching another monkey being
attacked may be fearing for itself as a possible victim, rather than
(or as well as) fearing for the other monkey. The clearest evidence
of animals frightened for others, as one might expect, comes from
parents frightened for their young.

Wildlife biologist Thomas Bledsoe describes the actions of
Red Collar, a mother grizzly brown bear whose cubs vanished
while she fished for salmon in the McNeil River, a gathering place
for bears. First she looked up and down the riverbank, then ran to
the top of the bluff and looked there, running faster and faster. She
stood on her hind legs to see farther, jerking her head around,
panting and drooling. After some minutes Red Collar gave up the
search and went back to fishing. Here her behavior is puzzling and

susceptible to varying interpretations, ranging from loss of interest (which humans would find difficult to identify with) to belief that no disaster had befallen her cubs. It is worth noting that on the occasions when Red Collar's cubs disappeared from the river, they had invariably gone off with one of the other mother bears and her family and were in fact safe. According to Bledsoe, at one point two of Red Collar's cubs were with another bear for three days before she encountered and reclaimed them.

Parents fear not only losing their young but also that they will be injured. Another brown bear Bledsoe observed, Big Mama, was alarmed when her two curious yearling cubs chose to investigate human observers, going after them uttering alarm calls until her cubs left the humans alone. Lynn Rogers, who studies the smaller black bear, says that when faced by danger, mother bears not only urge cubs up trees, but also discourage them from climbing smooth-barked trees like aspens in favor of rougher-barked pines (which are easier for small cubs to climb). Paul Leyhausen observed several mother cats who would allow their kittens to chase mice, but would interfere if the kittens went after rats. Tested away from their mothers, the kittens proved quite capable of tackling rats.

Mountain goat nannies vigilantly try to prevent their kids from taking dangerous or fatal falls. According to Douglas Chadwick, nannies try to stay on the downhill side of their kids, both when the kids are moving about and when they sleep. Due to the exuberance of the kids, the nannies must watch constantly. Chadwick notes of one mountain goat, "I could hear her literally cry out when the baby took a hard spill, and she would rush over to lick and nuzzle it, and then encourage it to nurse." The mother's cry is very like a human reaction to seeing someone fall, and it tells a perfect story of empathy.

A peregrine falcon father attacked one of his sons every time the young falcon came too close to human observers. Eventually the young bird changed his behavior, avoiding the observers afterward. The father's fear for his son altered the young bird's actions.

Social animals may fear for other members of the group. One experimenter decided to investigate the reaction of some young chimpanzees to "a bold man" and "a timid man." The chimpanzee Lia avoided the bold man, but the chimpanzee Mimi fought him. One day the "bold man" bent Mimi's finger back until she screamed. Lia joined the attack, but stopped when she got punched (such is the elegance of experimental research). After that, Lia devoted her efforts to trying to hold Mimi back, by grabbing her hands and pulling her away. In a group of caged chimpanzees at Oklahoma's Institute for Primate Studies, a female chimpanzee with an infant, whose previous babies had been removed, became apprehensive when approached by scientists. So did the other chimpanzees in the group in nearby cages. In this case, however, it is not clear whether the other chimpanzees were actually fearful; they may have been merely hostile. The deep fears that being in a laboratory may occasion in an animal have never been the object of study. Possibly the ethical dilemma created by causing such fear is too transparent to be acknowledged by scientific scrutiny.

The Spectrum of Fear

Fear at its mildest—a readiness to fear—may be characterized as caution or alertness and has obvious survival value. The alert worm hears the early bird coming and escapes. When this feeling intensifies it becomes anxiety, a painful uneasiness of mind. Psychiatry has made a good living from the fact that some people seem to be incapacitated by the degree of anxiety they feel, while others think their anxiety unnecessary or exaggerated.

Very great fear, like very great pain, can produce shock. The term *shock* has a medical definition, and there is no doubt that animals experience it. Hans Kruuk describes what looks like shock in wildebeest cornered by hyenas. These animals scarcely try to defend themselves once they have been brought to a standstill. They will stand in one spot, moaning, and be torn apart by the hyenas.

Pandora, a two-year-old mountain goat to be fitted with a radio collar, was trapped at a salt lick by wildlife biologist Douglas Chadwick and his wife. At first she made spirited attempts to escape. She tried to jump out of the enclosure, hooked a horn at Chadwick, and when tackled and brought down, attempted to get up again. While being blindfolded she went into shock, falling limp. Pandora had injured herself only slightly in this struggle, so it seems the reaction was the result of her intense fear. (After being collared, she was revived with smelling salts and released, showing no ill effects.)

In Africa a buffalo was knocked down, but not injured, by a lion, and simply lay on the ground in shock while the lion (perhaps an inexperienced animal) chewed on the buffalo's tail. Such an instance is another demonstration that fear does not always lead to survival.

Brave as a Lion

Bravery, sometimes considered an emotion, is related to fearfulness. Unfortunately bravery, or courage, is poorly defined in humans, so it is difficult to look for it in animals. It is often considered to involve proceeding against fear, overriding it or setting it aside. But is a dangerous act brave if you have no fear when you do it? Or is it only brave if you are afraid?

Hans Kruuk reports several instances in which cow and calf wildebeest were pursued by hyenas. In each case, when the hyenas caught up with the calf, the mother turned and attacked the hyenas, butting them so fiercely as to bowl them over. Perhaps this counts as bravery. Without a calf, a wildebeest cow keeps running. Surely fear makes her run. On the other hand, a human in a parallel situation may declare, "I was so angry I forgot to be scared." Perhaps a mother wildebeest is so angry she forgets fear. Is she brave?

For a nature program about cheetahs on television, a lioness was filmed killing a litter of cheetah cubs. While she was still there, the mother cheetah returned. Seeing the lion, the cheetah circled, hesitated, and then darted close to the lion until the lion pursued her. The cubs had already been killed, though the cheetah probably didn't know this. The mother cheetah evidently feared that the lion would kill her cubs and also feared being attacked by the (much larger) lion. Her attempt to draw the lion away seems to qualify as a brave act. After the lion was gone, the cheetah found the dead cubs, picked one up, and carried it away. During a sudden storm she was filmed sitting in the rain, crouched over the cub's body. When the rain stopped, she trotted off without a backward look.

Charles Darwin, too, was interested in animal bravery and gave the following account:

> Several years ago a keeper at the Zoological Gardens showed me some deep and scarcely healed wounds on the nape of his own neck, inflicted on him, whilst kneeling on the floor, by a fierce baboon. The little American monkey, who was a warm friend of this keeper, lived in the same compartment, and was dreadfully afraid of the great baboon. Nevertheless, as soon as he saw his friend in peril, he rushed to the rescue, and by screams and bites so distracted the baboon that the man was able to escape, after, the surgeons thought, running great risk to his life.

For Darwin, then, it was clear that a "mere" monkey could be a friend and a brave one at that. For this he was severely criticized by a modern scientist for his "tendency to anthropomorphize animal behavior," noting "it is small wonder that he was able to find evidence of all the human attributes [in animals], even moral behavior and bravery." It apparently upsets some scientists deeply for Darwin of all people to tell a story about the bravery of a small monkey who puts his own genetic future at risk for the sake of a member of another species, with whom he developed not dependence, but warm friendship. *Bravery* and *courage* are not words scientists are eager to see applied to a monkey by the founder of evolutionary theory.

Elephant calves, like mountain goat kids or bear cubs, do not always fear what their elders think they should. Cynthia Moss, who studies elephants in Kenya, reports that very young calves appear largely fearless. They may come up to her Land Rover and examine it—while people are in it. This often alarms their mothers and aunts, producing visible conflict, Moss reports. Apparently they would like to hustle the calves away but are too frightened to come close enough to do so. They stand very tall, and shuffle back and forth or swing one leg. When the calf eventually wanders back, the adults pull it to them, feel it, and make threatening gestures at the vehicle.

A Possible Need to Fear

By the time elephant calves are grown, they are likely to have had reason to fear things besides people in Land Rovers. Objects that justify fear appear in most creatures' lives. But what about an animal that is so protected and sheltered that it never encounters anything frightening? What happens to its capacity for fear? It is possible that such a creature will feel fearful anyway, that its capacity to fear will demand expression, fastening on an object that seems arbitrary.

Koko, a gorilla, was born in a zoo and raised by humans in a sheltered, loving environment. Koko was never exposed to big older gorillas, to leopards, to hunters, or to anything that might frighten her. Yet she has fears—of alligators, for instance, though she has never seen a real one. For years she acted afraid of toy alligators unless their lower jaws were missing. Though not frightened of her alligator puppet, she would play chasing games with it. She once threatened an aide in American Sign Language with being chased by an alligator if she didn't make lunch faster. She also appeared to be afraid of iguanas, specifically a pet iguana whom she saw often. Although the iguana (described as "comatose") never made threatening moves toward her, Koko would run into her own room if the iguana was brought out.

Possibly Koko's fear of lizards and alligators is instinctive, or partly instinctive; perhaps it is strengthened by the lack of anything else to fear. It may be that fear demands an object, and that no matter how secure and protected a child is, vampires, werewolves, or fire engines will be conjured up to serve as that object. In later years, perhaps because she received gifts of dozens of toy alligators of various descriptions, Koko seemed to lose her fear of them.

The chimpanzee Viki, raised by humans, had a fear of tarpaulins so severe that she could be kept from entering forbidden rooms by hanging pieces of tarpaulin on the doorknobs. The famous Washoe, while unimpressed by tarpaulins, has been reported to fear dust mops. Moja, another chimpanzee in the same group, was unmoved by dust mops but found the dividers from ice-cube trays so alarming that researchers kept ice-cube dividers hidden in drawers and cupboards so that, if Moja became unruly, they could punish her by taking out a divider and exhibiting it.

In a remarkable use of narration, Washoe and the other chimpanzees in her group were also led to fear an imaginary "bogeydog." This grew from an effort to get the easygoing young Washoe to use the sign for *no* more often. One evening researcher Roger Fouts looked out the window of Washoe's trailer, and signed to Washoe that he saw a big black dog with long teeth that ate baby chimpanzees. He asked Washoe if she wanted to go out, and got a most emphatic "no." On other occasions, when Washoe was playing outside and did not want to go in, researchers would sign that they saw the big black dog coming—and Washoe would hasten inside.

Island Fearlessness

Fearless animals sometimes greet travelers to remote islands. These bold creatures, rather than running from humans, may gaze attentively at a person walking up to them with a net or a gun. The botanist Sherwin Carlquist describes an encounter with an insular species of burrowing owl. He came close and photographed the owl, which stood and blinked lazily. A snake slithered by and Carlquist picked it up. Unperturbed, the snake draped itself on his shoulders and allowed him to carry it about all day. In the same archipelago, Carlquist was able to stroke elephant seals lying on the beach and boobies sitting on their eggs. Elsewhere, an island species of chuckwalla (a large lizard) was so placid that it would even disregard "a mild kick from a biologist's shoe."

Biologists who arrived on one uninhabited island after an arduous trip paused to rest; one researcher lying on the beach fell asleep. An island wren alighted on his foot, scrutinized his bootlaces, hopped along his body, perched on his chin and—to the joy of the sleeper's companions—peered long and carefully down each of his nostrils before flying off. Such fearlessness is found in species living on small islands where predators are few or absent. Carlquist argues, "Excessive skittishness is no virtue, in terms of evolution. If a bird spends much of its time flying away from false alarms, it will have that much less time for feeding and other essential activities. Thus, in a predator-free situation, a reasonably oblivious animal might be more successful than a perpetually nervous one." It is not reported whether island-tame species retain other fears—of heights or of water—but it seems likely. Many island species have calmly met their doom through fearlessness. The great auk and the dodo are only two who disappeared because they did not flee hungry humans or their animal companions.

The Other Side of Fear

If fear is the feeling that something bad is imminent, then its converse may be hope, the feeling that something good is imminent. In humans hope, like fear, can be unreasoning and irrational or logical and conscious. One of the most endearing traits of pet animals is their (quite reasonable) hope of being fed, and their unsophisticated joy at the prospect. Dogs whirl around in anticipation, cats purr loudly and rub against objects, people, or other animals.

When Washoe grew older, she had a baby that died four hours after birth because of a defective heart. Three years later she had a second baby, Sequoyah. Sequoyah was sickly, and despite excellent care from Washoe, died of pneumonia at the age of two months. Determined that Washoe should raise a baby, researchers made frantic efforts to find a replacement, and eventually procured Loulis, a ten-month-old chimpanzee. Fifteen days after Sequoyah's death, Fouts went to Washoe's enclosure and signed "I have baby for you." Every hair on Washoe's body stood on end. She displayed signs of great excitement, hooting, swaggering bipedally and signing "baby" repeatedly. "Then when she signed 'my baby,' I knew we were in trouble," Fouts said.

When Fouts returned with Loulis, Washoe's excitement vanished instantly. Her hair flattened and she declined to pick Loulis up, impassively signing "baby." But after an hour had passed, Washoe began approaching Loulis, trying to play with him. That evening, she tried to get him to sleep in her arms, as Sequoyah had done. At first she was unsuccessful, but by the next morning they were clasped together, and from that time Washoe has been a devoted mother to Loulis, who eventually acquired a vocabulary of fifty signs from Washoe and the other chimpanzees in the group. It seems clear that when told she would get a baby, Washoe hoped to see Sequoyah again.

Ludwig Wittgenstein believed that animals may feel frightened but not hopeful. He wrote, in the 1940s, "One can imagine an animal angry, frightened, unhappy, happy, startled. But hopeful? . . . A dog believes his master is at the door. But can he also believe his master will come the day-after-tomorrow?" Wittgenstein argues that only those who have mastered the use of language can hope. Not only does this statement remain unproven to this day, but there seems no good reason to doubt that an animal can imagine or even possibly dream about the future. Animals may lack the language of hope, but the feelings that underlie it are probably shared by humans and animals alike. If animals can remember and dream about the past, if fear can be relived, why can they not imagine and project a future in which fear will be unnecessary?

Love and Friendship

One evening in the 1930s Ma Shwe, a work elephant, and her three-month-old calf were trapped in rising floodwaters in the Upper Taungdwin River in Burma. Elephant handlers rushed to the river when they heard the calf screaming but could do nothing to help, for the steep banks were twelve to fifteen feet high. Ma Shwe's feet were still on the river bottom, but her calf was floating. Ma Shwe held the baby against her body; whenever she began to drift away, she used her trunk to pull the calf back against the current. The fast-rising water soon washed the calf away and Ma Shwe plunged downstream for fifty yards and retrieved it. She pinned her calf against the bank with her head, then lifted it in her trunk, reared up on her hind legs, and placed it on a rocky ledge five feet above the water. Ma Shwe then fell back into the torrent and disappeared downstream.

The elephant handlers turned their attention to the calf, which could barely fit on the narrow ledge where it stood shivering, eight feet below. Half an hour later, J. H. Williams, the British manager of the elephant camp, was peering down at the calf wondering how to rescue her when he heard "the grandest sounds of a mother's love I can remember. Ma Shwe had crossed the river and got up the bank and was making her way back as fast as she could, calling the whole time—a defiant roar, but to her calf it was music. The two little ears, like little maps of India, were cocked forward listening to the only sounds that mattered, the call of her mother." When Ma Shwe saw her calf, safe on the other side of the river, her call changed to the rumble that elephants typically make when pleased. The two elephants were left where they were. By morning Ma Shwe had crossed the river, no longer in flood, and the calf was off the ledge.

Too Noble for Animals

Humans—who are, after all, social primates—believe they know what love is and esteem it highly. Yet love is not considered an emotion by many theorists, but rather a "drive," like hunger.

Whether it is called emotion or drive, in most scientific circles it is forbidden to say that animals love. Had Williams been an animal behaviorist describing Ma Shwe and her calf, he would probably have felt constrained from using the word *love* to describe her behavior. He might instead have written of a "bond" between Ma Shwe and her calf. In a critique of Harry Harlow's deprivation experiments in which young rhesus monkeys were forced to grow up without mothers, biologist Catherine Roberts has written, "Does he not know that human love differs qualitatively from animal love? Does he not know that a human mother is unique because she has an abstract idea of the Good and that therefore human love, unlike animal love, has its ontogenetic beginnings in a spiritual bond between mother and child?" Animals, in other words, cannot love like people can because their bonds are not spiritual.

The evolutionary approach stresses the survival value of love over its emotional authenticity. The writer of one book for a popular audience has noted that there is general approval of animals that mate for life, and added, "It's important to remember . . . that these animals are not displaying 'true love,' but simply following the dictates of their genes. They are survival machines, and their mission is to multiply their own genes in the gene pool. If a male felt that his partner could raise young without him, he'd be off in a flash. But this would not be abandonment in our terms, and we need not feel sorry for the female. Both are pursuing their own game plan that will lead to the best positioning of their genes—a pursuit that is adaptive and, therefore, beautiful." Whatever their scientific view of human coupling may be, most people would not accept this as an accurate way of looking at their own loves and families, yet where the difference on this point lies between animals and people is not stated. Would the most "beautiful" human love be one that sought dominant genes to reproduce? The fact that an animal is described on the one hand as a machine, and on the other as a being that may ponder whether its partner can raise young alone is only one of the inconsistencies in such thinking. Ostensibly objective statements like this that reduce the complexity of inner life to its function are all too typical. Perhaps love, the emotion, has evolutionary value. Elizabeth Marshall Thomas argues of two dogs, Misha and Maria:

> Popular prejudice might hold that romantic love, with its resulting benefit of fidelity, sexual and otherwise, is not a concept that can be applied to dogs, and that to do so is anthropomorphic. Not true. Fully as much as any human love story, the story of Misha and Maria shows the evolutionary value of romantic love. The force that drove Romeo and Juliet is no less strong or important if harbored by a nonhuman species, because the strength of the bond helps to assure the male that he, instead of, say, Tybalt or Bingo, is the father of any children born and that both parents are in a cooperative frame of mind when the time comes to raise those children.

Even though she observed that the emotion could serve a scientific rationale, Thomas has been castigated for using the word *love* rather than *bond* in writing of dogs.

Parental Love

The evolutionary approach suggests that parental love—watching over young—makes urgent sense. Parental care allows more young to survive. If parents protect their young, the young can grow bigger before they have to fend for themselves. A baboon can even inherit its mother's status in the troop and an adult female black bear can use her mother's territory while her mother is still occupying it. A young animal can learn survival practices while safely under the protection of its parent. Perhaps—this is debated—the parent even teaches it some of those things.

Not all creatures protect their young. A turtle lays eggs in the sand and departs. Presumably it would not recognize, let alone love, its offspring. But if an animal lays eggs and guards them, as crocodiles do, there must be something that motivates it to do so, and then prevents it from eating the young when they hatch. This might not necessarily be love—it could be brought about by such simple mechanisms as an inhibition against eating eggs and young crocodiles. But expressing care may be evidence of feeling love. Crocodiles also dig their young out of the nest when they hatch, guard the babies, carry them in their jaws, and respond vigorously to their distress calls. Females of a southeast Asian diadem butterfly apparently guard their eggs by standing over them. This probably increases their chances of survival. However, a female will sometimes continue this behavior even unto death, her rotted corpse standing guard over a batch of as yet unhatched eggs.

Mother wolf spiders not only tend their eggs but carry their babies on their backs. Perhaps the babies need to learn hunting skills. More likely, they just need protection while they grow. J. T. Moggridge tells the story of a trap-door spider he had collected and decided to preserve in alcohol. While he knew that spiders twitched for a long time after being put in alcohol, it was then believed that this was mere reflex action. Moggridge shook the baby spiders off her back and dropped her into alcohol. After a while, supposing her to be "dead to sense," he dropped her twenty-four babies in too. To his horror, the mother spider reached out her legs, folded the babies beneath her, and clasped them until she died. After this, Moggridge switched to the use of chloroform.

Can a spider love her babies? Was it a mere reflex that caused the trap-door spider to reach for her young? In this case it seems possible, but it is hard to be certain. One can imagine a simple instinct to draw close to anything that looks like a baby spider. Or she might have seized any objects that happened to be floating in the alcohol. A mother wolf spider is just as kind to strange baby wolf spiders as to her own. This might or might not be accompanied by an emotional state.

Does a spider love its eggs, something the writer John Crompton compares to loving a box of billiard balls? It is so hard to have insight into a spider's mind that it is almost impossible to guess, based on present knowledge. Yet spiders have evolved to produce complex venoms and digestive fluids, and spin silks of varying types from six different kinds of silk glands. Building a spider's web is an extremely complicated behavior. One can argue that a spider is not really a simple organism and that the development of maternal love might well be a shorter evolutionary step than web building. Perhaps one day we will know. What if it was discovered that when a mother wolf spider sees young spiders, her body is flooded with a hormone whose presence is associated with feelings of love in higher animals? Would that be evidence that the spider loves her young? What if it was a hormone peculiar to spiders? Would that mean it wasn't love?

When trying to comprehend the inner lives of creatures so unlike us, it is more useful and accurate to think not of a hierarchy with human beings at the top, but of a spectrum of creature commonality. A spider might have a rich inner life with a riot of emotions including some so different that using our own emotional range as a touchstone can only fail us.

While the question of whether a spider can feel parental love is baffling, there seems little doubt for "higher" animals. Their behavior is so complex that to dismiss it as the exclusive result of inhibitions, reflexes, and fixed action patterns is patently inadequate. Parental care manifests itself in feeding the young, washing them, playing with them, and protecting them from external dangers and from their own inexperience. Mammals, even "primitive" ones such as platypuses and spiny anteaters, suckle their young. The suckling mother is extremely vulnerable in a way that she will seldom allow herself to be with adult animals—in a way that many protective instincts would advise against.

Young mammals are safest in their own nests. In a series of classic experiments on rats, researchers put baby rats on cage floors. Mother rats, and in some cases females who were not mothers, proved zealous at retrieving the babies and bringing them into their nests. They would cross an electrified grid to get to the babies and retrieve unrelated babies as quickly as their own. Curious to see how long this would be kept up, the experimenters offered one rat no fewer than fifty-eight babies, every one of whom she picked up and crammed into her nest. "The female appeared to be as eager at the end of the experiment, which had to be interrupted because we had no more young at our disposal, as she had at the beginning." This behavior did not enhance her own survival. Similarly, when biologists climb up to ledges to band young thick-billed murres (penguinlike seabirds), most of the adults fly away in panic, but a few staunch birds sit tight. Frightened chicks whose own parents have flown off seek out the remaining adults. "It is not uncommon to see one motivated brooder vainly attempting to shelter a dozen or more chicks," seabird biologists have noted.

In contrast to the zealous rats and motivated murres, Nubian ibex who bear triplets instead of twins are reported to reject one fawn. Presumably the doe cannot produce milk for three, so if she kept them all, they would all be malnourished. This "lifeboat behavior" could also be a form of ethically responsible love. When most of a lioness's cubs die or are killed, she may abandon the last cub. Some biologists suggest that it is energetically inefficient for her to put the effort of raising a litter into just one cub, when she can breed again sooner if she does not, and that her "instinctive sense of investment" tells her so. What an instinctive sense of investment may be or feel like, and how it differs from making difficult, loving decisions under conditions of constraint, is not clear. Human parents have been known to take similar actions.

What goes on in the mind of an ibex doe or a lioness in this situation is as yet unknown.

As young mammals grow older, their parents often feed them. Some animals simply let the young steal pieces of their own food, while others bring food to the young. An ocean bird may begin by regurgitating partially digested food for its chick. When the chick is older, the parent shifts to bringing whole fish, which it holds until the chick manages to grab it properly. As baby animals grow bigger, most of them play, sometimes with their littermates and sometimes with their parents. As anyone who has observed kittens or puppies knows, young animals can be rough on adults. Biologists studying wild dogs in Africa commented that when the dogs brought food to a mother and her three-week-old puppies, the pups were aggressive in taking their share. If their mother had a chunk of meat a pup wanted, it would tweak the side of her face with its sharp teeth and she would let go. Older pups followed hunting adults and took over the carcasses of prey, sometimes even nipping the adults on the rear to make them leave faster.

One of the most important tasks of parents is to protect the young. Baby animals are typically small, inept, and defenseless and make desirable meals for predators. Some parents protect their young by hiding them. At other times parents must fight to save their young. In a typical instance, a lion was seen to attack a herd of six giraffe. Most ran away, but one calf was too slow. Its mother tried to push the calf to run faster, but when she saw this would not work, she stood over it and faced the lion. The mother was in real danger, since lions frequently succeed in killing giraffe. The lion circled the giraffe, and the mother wheeled to face him. Whenever he got close, she kicked at him with her forelegs. After an hour, the lion gave up and left. The two giraffe rejoined the herd.

The willingness of animal parents to fight in defense of their young is well-known, since it is often humans who are the threat. But humans are so menacing that such encounters seldom develop into actual fights. The last known whooping crane nest in the United States was found in Iowa by ornithologist and egg collector J. W. Preston. He wrote, "When I approached the nest, the bird, which had walked some distance away, came running back . . .

wings and tail spread drooping, with head and shoulders brought level with the water; then it began picking up bunches of moss and sticks which it threw down in a defiant way; then, with pitiable mien, it spread itself upon the water and begged me to leave its treasure, which, in a heartless manner, I did not do." This might have been either a male or female crane, since both sit on the eggs.

Animals also try to rescue their young from dangers other than predators, as in the case of a cat who had never entered water but jumped into a swimming pool to rescue her kittens. To the north of Hudson Bay, explorer Peter Freuchen came across a family of six wolves, two adults and four cubs. The wolves were howling. One of the cubs was caught in a trap set at a cairn of stones over a food cache. The other wolves had overturned many of the large stones and scraped the frozen earth around the stone to which the trap was fastened in their efforts to free the cub. When humans protect their young this way we call it love.

Since so many examples given of parental love are of motherly love, it is worth stressing that fatherly love is apparent in some species. It is estimated that direct paternal care is found in ten per cent of mammalian genera. Paternal care ranges from slight to intense devotion. The conservationist Gerald Durrell has described the birth of cotton-topped marmosets at the Jersey Zoo. After the mother had delivered the usual marmoset twins, the father took them, washed them, and carried them with him everywhere he went, often one on each hip, only returning them to the mother to suckle. As they grew older, they would leave his side to explore. If he felt there was a threat to their safety, he would rush over and snatch them up. Wild marmoset fathers of various species behave in the same way. Often they assist at the birth. Lion-headed marmoset fathers have been seen to mash fruit in their fingers for the babies when they begin to wean. Father owl monkeys (or night monkeys) also usually carry their babies, play with them, and share food with them. As a result, a father owl monkey often trails behind the mother and older offspring and arrives at fruit trees when they are already partly depleted. The mother suckles the infant and then returns it to the father.

Red fox researcher David Macdonald describes a new father wriggling with eagerness to care for his children:

Smudge was almost comical in his husbandly diligence. Before eating a scrap for himself, he gathered as much food as he could wedge into his gaping jaws, and lugged it to Whitepaws' earth. There he would warble at the entrance. If she did not emerge, he would use his nose like a billiard cue to poke the lumps of food through the entrance and into the den.

When the cubs were older, Smudge's ambition was to play with them, something the mother and her sisters did not always allow. "Smudge would skulk in the vegetation, waiting for Big Ears [their maternal aunt] to fall asleep, whereupon he would quietly warble to the cubs who would sneak off to gambol with him. Soon their exuberance led to squeals and snirks that awoke Big Ears who would vigorously reprimand Smudge."

The psychoanalytic paradigm of murderous rivalry between fathers and sons claims to represent the state of nature. Psychoanalysis, like sociobiology, has been mainly interested in observations that support received theory. Counterexamples tend to be ignored: father zebras remain on good terms with their grown sons, who eventually leave the herd not because they are driven out but because they are looking for others to play with. Researchers studying wild zebras once decided to mark a stallion who was still living in his father's herd at the age of four and a half years, so that they could follow his subsequent travels. To their sorrow, the anesthetic dart killed him. The old stallion came over to the body repeatedly and tried to rouse him. Later that day he spent hours roaming from herd to herd, calling for his son.

Direct paternal care is seen in many species of birds, including the kiwi, in which case the father incubates the eggs and raises the chicks without help from the mother. In many other instances— the father beaver playing with his children; the father wolf letting his cubs chew his tail; the father dwarf mongoose taking the kits foraging with him—it appears that they love their children, or at least enjoy their company.

Is It Love?

In these and other ways animal parents appear to act out of love for their young. To argue that it cannot be compared to human love, as many theorists do, is a classic example of what Roger Fouts calls the rubber ruler, in which the standards are changed depending on whether the behavior is human or nonhuman. In considering if it is possible to know whether a mother ape loves her baby, it is worth asking if it is possible to know whether the people down the street love their baby. They feed it and care for it. They tickle it and play with it. They defend it with all their might. But all that is not considered proof in the case of the ape.

Unlike the ape, the people down the street may *say* they love their baby, but how do we know they are telling the truth? Ultimately we cannot know exactly what other people mean when they speak of love. Yet in reality, we are usually quite sure they love that baby. If we see parents with a baby they don't love, we are shocked. When we learn of child abuse, we are outraged, in part because we believe that a relationship of love is violated. In truth, most people believe that the ape loves her baby, that the dog loves her puppies, and that the cat loves her kittens. Most scientists probably believe it, too, though they may be hesitant to say so, at least in a scientific document. A skeptical observer might still object that an ape with her baby is acting out of mere instinct. Could it not be that the people down the street are acting out of instinct as well? It depends on whether love is defined as an instinct. And, in either case, if it is called instinct, does that mean no feelings of love can also be there?

On the flip side of parental love is filial love, the youngster's love for his or her parents, which is harder to pin down. To almost any demonstration of attachment on the part of an animal to its parent—of the tiger cubs licking their mother, of the wolf cubs running to greet their father—the skeptic can simply argue that it is self-interest. The younger animal may simply wish to be around the source of food, warmth, and safety. Young animals do not, as a rule, fight to protect their parents. Nonetheless, one adolescent wild baboon, Paul, tried to defend his mother against large adult males in the troop. He was not very successful, but a scientific observer felt that he risked a great deal in his mother's defense. When they grow older, many young animals are so reluctant to leave their family that they must be driven away by their parents. But this alone cannot prove that they love their parents. They might be reluctant to depart from safety and their accustomed habits, to go into unknown territory. Not all young animals are driven away—in some families the bonds of affection are lasting. Chimpanzees usually remain in the same group as their mothers. They spend time with each other, and the young apes may help their mothers care for the next generation. Elephants, too, live in stable maternal herds and cooperate in extraordinary ways with one another. Elephant aunts play important roles in child care.

In a captive baboon colony, experimenters removed female baboons as soon as they began to have menstrual cycles after the birth of an infant. The babies were usually about six months old and were left behind. Other females in the colony took over their care. Seven or eight months later, the original mothers were returned to the colony, often anesthetized. Each time an unconscious baboon was carried in, her baby began to utter "lost baby" calls. When she was returned to the enclosure the baby went to her and the mother-child relationship resumed. Despite the excellent care they were receiving, these young apes recognized and yearned for their mothers. Baboons are individuals, however, and one baby chose to remain with its foster mother.

Jane Goodall describes the reaction of a male chimpanzee called Flint, who was eight years old when his mother, Flo, died. Flint sat over Flo's body for many hours, occasionally tugging at her hand. As the days passed, he grew increasingly apathetic and lethargic. In one remarkable instance, three days after his mother's death, Flint was seen to climb a tree and stare at the sleeping nest he had shared with his mother a few days before. He became more and more listless, and died within the month, probably of gastroenteritis. Goodall's scientific conclusion was: "It seems likely that psychological and physiological disturbances associated with loss made him more vulnerable to disease." Sy Montgomery quotes Goodall's poignant commonsense rendering of the same idea: "Flint died of grief."

Adoption

The flexibility of parental love is seen in those animals who adopt unrelated babies. Black bear researchers regularly persuade wild females to adopt orphaned cubs. Researchers in Africa kidnapped infant and juvenile hamadryas baboons and released them near unrelated troops. Invariably these young monkeys were promptly adopted by young adult male baboons, who cared for them tenderly. In general, the younger a baby animal is, the more apt it is to be adopted.

Even greater flexibility in parental love is demonstrated by instances of animals adopting the young of other species. The experimenters who gave the mother rat the opportunity to adopt fifty-eight babies went on to offer mother rats odder babies. The rats readily adopted baby mice and baby rabbits. They also retrieved young kittens and tried to keep experimenters from taking them out of the nest again. But since kittens nurse from a mother who is lying down, and mother rats nurse from a standing position, the rats could not suckle the kittens despite their vigorous attempts to shove them into position. Curious to know how far this would go, the experimenters procured two bantam chicks, and the rats "eagerly and repeatedly" tried to tuck these into their nests. This was an even worse match, however, as the chicks "became very loud and flapped" when the rats tried to grab them by the neck and drag them home.

There are, of course, innumerable cases of dogs or cats who have adopted orphaned skunks or piglets, and pictures of such odd families are frequently seen in the newspapers. There is an old story of an unusual adoption, not initiated by humans:

> At Northrepps Hall, near Cromer, the seat of the late Sir Fowell Buxton, a large colony of parrots and macaws had been established, for whom a home had been provided near the house in a large open aviary, with hutches for them to lay in. But the birds as a rule preferred the woods . . . only coming home at feeding-time, when, on the well-known tinkling of the spoon on the tin containing their food, a large covey of

gayly plumaged birds came fluttering down to the feed-place, presenting a sight not often to be seen in England. The hutches being then practically deserted, a cat found one of them a convenient place to kitten in. While the mother-cat was away foraging, one of the female parrots paid a chance visit to the place, and finding the young kittens in her nest, at once adopted them as her own, and was found by Lady Buxton's man covering her strange adopted children with her wings.

Other species seem not to adopt. According to researchers, a wildebeest calf who loses its mother among the vast herds will not be adopted and will perish. This raises the question of how selectivity in love is esteemed. Humans are not likely to be favorably impressed when parents cannot tell whether or not a baby is their own, but they also do not like to see parents discriminating too harshly against the babies of others. There seems to be no reason why either response should be considered incompatible with parental love, though we esteem one more than the other.

Imprinting

Ducklings and gosling will grow attached to and follow whatever creature they see during a short period after they hatch. They become "imprinted" on that creature. In this sense, a mallard is not born to love only a mallard, and a teal is not born to love only a teal. They come to love whomever seems to be their primary caretaker. For a mallard this is usually a mallard, and for a teal a teal, but mallards raised by humans, and who become attached to their human foster parents are common. It is often said of animals raised by humans, "He thinks he *is* a human." Or maybe he thinks the human is a mallard. Either illustrates the flexibility built into filial love.

Besides parents and children, love can extend to other family members. One young wild elephant seemed to be as fond of his grandmother, Teresia, as he was of his mother. Often he would suckle from his mother and then go to Teresia, who was more than fifty years old, and stand with her or follow her. In many species, from beavers to gibbons, young animals may remain with their parents and help raise the younger offspring. Young coyotes often remain with their parents and help raise subsequent new pups. These older siblings may feed, wash, protect, and simply baby-sit the pups. The arrangement, referred to as allomothering, clearly benefits the parents, who need all the help they can get with raising the pups; and it benefits the pups, who have more adults looking after their welfare. In some instances the helpers are siblings of the parents, not of the young. If the parents are killed, the older siblings or aunts and uncles can raise the pups if they are not too young. In one pack of wild dogs in the African savanna, the mother of nine pups died when they were five weeks old. Apparently they were old enough to switch to solid food, for the rest of the pack, consisting of five males, reared them successfully.

Avoiding any mention of fondness, evolutionary biologists have identified a number of ways this benefits siblings and aunts and uncles. Possibly they have more time to learn hunting skills. If there are no good coyote territories available, they are spared the necessity of going out and fighting for one. They also increase the chances of their genes being passed on, since they share many genes with their young siblings, nieces and nephews. But possibly, also, they love their families.

Year-old beavers typically remain with their parents and help take care of their younger siblings. Françoise Patenaude, who observed wild beavers in Quebec, saw the yearlings grooming the babies and fetching food for them. During the winter, the entire family was largely confined to the lodge. On more than one occasion, when an infant beaver fell into the water in the entrance to the lodge, a yearling picked it up and carried it in its arms to the dry floor of the lodge. (Beavers can walk on their hind legs and carry things, including very small beavers, in their forelegs.) The yearlings later played with their younger siblings and helped perform every aspect of parental care except, of course, suckling.

Social Animals

Social animals who live in groups often behave in a friendly way toward other members of the group, even when they are not relatives. Troops of baboons and herds of zebra or elephants are not just crowds of strangers. This can go far beyond toleration to a kind of need: a monkey kept alone will work for the reward of seeing other monkeys, just as a hungry one will work to get food. The animals in a social group have relationships with each other, some of which are affectionate. Lionesses baby-sit for one another just as house cats sometimes do. In baboon troops, many baboons have alliances with other baboons, whom they can count on to take their side in squabbles.

Elephants appear to make allowances for other members of their herd. One African herd always traveled slowly because one of its members had never fully recovered from a broken leg suffered as a calf. A park warden reported coming across a herd with a female carrying a small calf several days dead, which she placed on the ground whenever she ate or drank: she traveled very slowly and the rest of the elephants waited for her. This suggests that animals, like people, act on feelings as such, rather than solely for purposes of survival. It suggests that the evolutionary approach is no more adequate to explain animal feelings than human ones. A single example such as this one, no matter how well documented, may not challenge the entire evolutionary paradigm for feelings, but it raises questions that biologists have yet to face. There appears to be so little survival value in the behavior of this herd, that perhaps one has to believe that they behaved this way just because they *loved* their grieving friend who loved her dead baby, and wanted to support her.

As with humans, sometimes affection among animals combines with admiration, and can flow to other species as well. Wolves display what appears to be admiration for the dominant (or alpha) wolves in their pack. In the wolf's close relative, the dog, this capacity to admire leaders has made domestication a success. The average dog—in its eagerness to please—treats its owners the way a wolf treats an alpha wolf.

Other social animals are also often willing to accord humans this kind of status. Researcher Jennifer Zeligs, who studies and trains sea lions, appears to be in this position with her experimental subjects. Her success in training them to retrieve objects underwater and to perform other tasks seems to be the result of their desire to please her and receive her praise and attention. She does not reward them with food, working with them only after they have eaten. One could object that this is not love, because the sea lions get pleasure from Zeligs: attention, grooming, and fun. That is why they give her the attention they do. But is human love any different? Must love be unrewarding to be real? The point is that sea lions seem capable of love and affection. That they can, under special conditions, extend these feelings beyond their own kind, to an alien species, merely demonstrates some of the conditions of those feelings. Two sea lions can feel affection for one another, and Zeligs benefits from the extension of that feeling to a human.

Sometimes the benefit of social behavior appears obvious, but at other times the advantage is less clear. Hans Kruuk's study of spotted hyenas showed that their social behavior was highly advantageous, but was puzzled when he turned his attention to European badgers. Though they live together in communal burrows or setts, badgers do not forage together, patrol territory together, defend each other, or cooperate to raise their young. "Badgers do not have an effective alarm call, so they do not even warn clan members of approaching danger." The only material benefit Kruuk could point to was the practice of sleeping in a huddle to keep warm.

If it is true that social life offers badgers no survival edge, then why do they live together? Perhaps they simply enjoy the companionship.

Friendship

In general, animals are friendly only toward animals of their own species. Significant exceptions occur in captive animals, who are often isolated from others of their kind or confined with animals of other species. In these circumstances some animals make friends with animals of other species, including humans. John Teal, who experimented with raising endangered musk oxen, was once shut in a pen with them when some dogs came running up. To his alarm, the musk oxen snorted, stamped, and thundered toward him. Before he could move, they formed a defensive ring around him and lowered their horns, pointing at the dogs. This is how musk oxen protect their calves from predators. But friendship with an animal of another species does not guarantee friendship with the entire species: a hand-reared leopard was raised with a dog and loved to play with her, but tried to kill other dogs, even dogs that closely resembled her friend.

Although it is rare, wild animals have been observed in friendly associations. Biologist Michael Ghiglieri, patiently waiting for chimpanzees to come to a fruiting tree in the Tanzanian rain forest, was astonished when the first chimpanzee, a male, arrived in the company of an adult male baboon.

Overtures of friendship are not always received kindly. Wild dogs and hyenas in the Serengeti are competitors who regularly steal kills from one another. A group of wild dogs had had a kill stolen by spotted hyenas and then chased one hyena, biting its rump so fiercely that it sat in a hole and snarled until the dogs left. Yet that evening, when the wild dogs bedded down for the night and the hyenas prowled, a young hyena (still with a fluffy coat) approached the dominant male wild dog, Baskerville, sniffing him enthusiastically. Baskerville twitched and growled, yet every time he tried to go back to sleep, the hyena inched closer. The hyena began to lick and groom Baskerville, who at first seemed to be ignoring this. Observers could tell he was not asleep, however, for his eyes got wider and wider as the young hyena continued these sociable ministrations. Baskerville pulled himself into a tight ball and glared over his shoulder, but the hyena calmly lay down next to him, apparently ready to settle in for the night. This was too much for Baskerville. He leapt to his feet with a loud bark. The pack awoke and moved off—followed by seven spotted hyenas. Eventually the wild dogs were able to elude the hyenas, but when they killed a gazelle the next morning, hyenas stole it from them. Given the relationship between hyenas and wild dogs, it is not hard to understand why Baskerville rejected the young hyena's advances. Perhaps many friendly gestures between wild animals receive cold responses for similar reasons.

Sometimes animals try to make friends across species out of desperation. In Madagascar a brown lemur was trapped and transported to another area, where he escaped. There were no brown lemurs, so he joined a troop of ring-tailed lemurs. The species have different coloration, calls, scent glands, and marking habits. The ring-tails did not welcome the brown lemur, but tolerated him. The males deferred to him and so, by definition, he became the dominant male in the troop. The females would not allow him to scent-mark them and usually would not let him sit next to them and groom them. During one breeding season he made advances to a female named Grin, who let him sit with her only five times in twenty-four attempts and who refused to mate with him, although she mated with some male ring-tails. His friendliest reception came from juveniles, who not only let him sit with them, but groomed him, and occasionally made the first approach. While his acceptance into the troop was only partial, it clearly meant a great deal to him.

Even an animal not considered social may make friends in captivity. Ocelots, usually considered solitary animals, can become quite friendly with humans. Perplexed by the amicability of his ocelot, Paul Leyhausen speculated that such cats have a capacity to be friendly as kittens, but that once adult they cannot help seeing other cats as rivals or intruders. Humans, he hypothesized, are similar enough to be friends with, yet sufficiently dissimilar that we do not seem like rivals: "Thus genuine and lasting friendship, of a kind which may never occur between cats themselves, is possible between humans and members of various species of solitary cats. In other words, if the above hypothesis is correct, the individual wild cat would really 'like' to be friendly with other cats, but feels toward them much like the eccentric who offends everyone and then, asked why he has no friends, replies in amazement, 'I wish I did, but everyone else is so horrible!' "

It is rare for animals to make friends with humans when they are not captive, for their own species make more suitable friends, and humans are usually feared. Wild beavers, given time, will tolerate well-mannered humans. If the humans supply favored beaver food, the beavers will associate with them, even to the point of climbing into their laps to get especially tasty food. They distinguish humans they know from strangers. Yet there is no reason to suppose that the beavers actually enjoy the company of the humans. Lack of fear does not equal friendship.

When Animals Have Pets

Friendship between people and animals often occurs when the animal is a pet. Animals also have pets occasionally—usually captive animals, since having a pet is a luxury. Lucy, a chimpanzee reared by humans, was given a kitten to allay her loneliness. The first time she saw the little cat, her hair stood on end. Barking, she grabbed it, flung it to the ground, hitting out and trying to bite it. Their second encounter was similar, but at their third meeting she was calmer. As she wandered about the kitten followed her, and after half an hour Lucy picked it up, kissed it and hugged it, marking a complete change of attitude. Subsequently she groomed and cradled the kitten, carried it constantly, made nests for it and guarded it from humans. She behaved as do affectionate small children with imperfect ideas of what actually pleases their pets. The kitten "never appeared anxious to be transported by the chimpanzee" and was unwilling to cling to Lucy's stomach, so she either carried it in one hand or urged it to ride on her back. Koko the gorilla showed great tenderness toward a pet kitten she herself named All Ball. This is strikingly like love, the real thing, because it is chosen without regard for survival value.

It is common for horses to make friends with other animals such as goats. They are careful not to hurt the goats. Accounts of racehorses who mope and do not run well when separated from their goat friends are numerous. These goats may well be the equivalent of pets. The horses are not confused about species: they know that the goat is not a horse, but they like it anyway. It is also reported that one captive elephant routinely put aside a little of its grain for a mouse to eat.

Romantic Love

As much as people admire friendship and familial love, our highest esteem is accorded romantic love, which is considered the most suspect to ascribe to animals. Many people consider romantic love so rarefied that they believe people in other cultures do not experience it, let alone animals. It has been claimed that the whole idea of romantic love was invented in medieval Europe and that it is only a pastime of the privileged. In any case, anthropologists studying human beings had not deemed it a fit subject for study until the American Anthropological Association held its first session on the anthropology of romance in 1992. Anthropologist William Jankowiak said it took him three years to organize the session. "I would call up people and they would just laugh." When asked why romantic love had been ignored, Jankowiak said researchers had assumed that "this type of behavior was culture-specific." He also pointed to a bias toward linguistic evidence. "The dominant model was a linguistic one that said if it isn't in the language it's not important." Not only do some cultures lack words for romantic love, but it is undefined in the anthropological lexicon. "They themselves don't have the categories for this."

Jankowiak said that when he asked colleagues for contributions to the session on love and was told in reply that romantic love did not exist in the cultures they studied, he would ask whether people in those cultures ever had clandestine affairs, ever refused arranged marriages, ever eloped or committed suicide over love. The answer was always yes, such things did occur, but that the researchers had not examined it or followed up on it.

Another reason that romantic love in other cultures has been ignored is the assumption that it is a frill. As Charles Lindholm, a pioneer in this field, said, "The general paradigm in anthropology, as in all social sciences, is utilitarian, maximizing gain. Romantic love doesn't seem to fit the paradigm very well. . . . When you have people sacrificing their lives for each other, it doesn't seem like you're maximizing gain. . . ." He laughed. "Another part of it is that it's embarrassing! You're asking people about personal relationships and anthropologists, like everybody else, aren't comfortable asking people about that." Nor has the general academic milieu favored the study of love. "It's sort of a woman's type of thing, you know. It's not good for your career."

All these factors—the denigration of love as a luxury or feminine, the absence of linguistic evidence, the stress on utility, the embarrassment, the lack of a theoretical framework, even the career concerns—may help explain the parallel lack of the study of romance among animals.

Jane Goodall, whose work has illuminated the emotional life of chimpanzees, has nevertheless argued that chimpanzees do not know romantic love. She describes the chimpanzees Pooch and Figan, who repeatedly showed a preference for one another when Pooch was sexually active, leaving the group and going off together into the forest for a few days. This is in contrast to chimpanzees who remain with the larger group while sexually active. Goodall writes, "I cannot conceive of chimpanzees developing emotions, one for the other, comparable in any way to the tenderness, the protectiveness, tolerance, and spiritual exhilaration that are the hallmarks of human love in its truest and deepest sense. . . . The most the female chimpanzee can expect of her suitor is a brief courtship display, a sexual contact lasting at most half a minute, and, sometimes, a session of social grooming afterward. Not for them the romance, the mystery, the boundless joys of human love." Perhaps this is true. And yet we do not fully grasp what chimpanzees feel, or whether there are boundless joys of simian love chimpanzees feel that we do not.

Some animals pair for life, staying together for as long as both live. Some animals pair off for a season, and others mate and separate at once. Among animals who form relationships of significant duration, some form pairs and others form larger groups, such as trios, or the elephant seal "harem." Evolutionary biologists often describe pairing as a device to ensure adequate parental care, but it is not always clear that this is the case. The butterfly fish of Hawaiian reefs (Chaetodontidae), for example, are said to provide no parental care for their eggs or larvae but do form lasting pairs.

Some people argue that there is no affection between animals who do not form partnerships, who mate and then separate, but this is illogical. A. J. Magoun and P. Valkenburg, who used a small airplane to track wolverines across the tundra, have described the mating of these rare, solitary animals. To an observer, they write, most wolverine mating appears to be a matter of aggressive males and reluctant females. They were surprised by the behavior of the female they called F9 and an unidentified male. F9 and the male joined in exploring a rock outcropping on the tundra. They played. They rolled on the ground. Like an exuberant dog, F9 crouched and lashed her tail, then bounded away. When the male did not respond to her sniffing him, F9 turned and bumped him with her hip. After playing, they rested and then mated. Two days later they separated, perhaps never to meet again. This describes a friendly, playful interaction, not one-sided lust or a convenient arrangement. Is it reasonable to call it love? Liking? Mutual interest? F9 was just a year old, so her playfulness might be ascribed to youth. Yet even if her emotions—and those of the male—sprang from her youth, that does not mean they did not exist.

It might be argued that whatever one wolverine feels for another, it cannot be love because the encounter is so brief. But to use duration as a valid measure of love would rule out much love between humans.

> *After all, my erstwhile dear,*
> *My no longer cherished,*
> *Need we say it was not love*
> *Just because it perished?*

It is reasonable to say that F9 and her consort enjoyed one another's companionship for those few days, that they liked each other. There was no necessity to play together: they did so because they wanted to.

If it seems suspect to grant romance to wolverines and their brief affairs, what about the animals who do mate for life? These are animals who court, mate, raise young, and accompany each other when not raising young. Sometimes they are in a larger group of animals, such as a flock of swans. At other times they may leave the group, as when swans are actually nesting. Mates commonly sleep together, groom one another, and in some cases forage together. Usually they feed one another only when courting or when one is staying with very young offspring. In most species sexual activity is confined to a brief period.

The most common proof offered of love between mates is the sorrow they exhibit when one of the pair dies. Konrad Lorenz describes as a typical example the behavior of the gander Ado when his mate, Susanne-Elisabeth, was killed by a fox. He stood silently by her partly eaten body, which lay across their nest. In the following days he hunched his body and hung his head. His eyes became sunken. His status in the flock plummeted, since he did not have the heart to defend himself from the attacks of the other geese. A year later, Ado had pulled himself together and met another goose.

Animals may fall in love dramatically. Konrad Lorenz said that two greylag geese are most apt to fall in love when they have known each other as youngsters, been separated and then met again. He made the comparison to an astonished human asking, "Are *you* the same little girl I used to see running around in pigtails and braces?" and went on, "That's how I met my wife." According to parrot behavior consultant Mattie Sue Athan, it is common for some of the larger species of parrots to fall in love at first sight, and this is known as "the thunderbolt."

Animals don't fall in love with just anyone. Seeking a mate for a male umbrella cockatoo, Athan purchased a young female cockatoo with beautiful plumage and put them together. To Athan's chagrin, "He acted like she wasn't even in the room." A few months later Athan was given an older female in bad condition. She had been plucking her feathers out, an effect of captivity. "She didn't have a feather from the neck down. The skin on her feet was all gnarly. She had wrinkles around her beak. He thought she was the love of his life." The two birds immediately paired off and began rearing a series of baby cockatoos.

Zookeepers know to their despair that many species of animals will not breed with just any other animal of their species. Orang-utans are among the most notoriously selective, even though they do not form lasting pairs in the wild. No doubt wild animals are selective, too, but this is less obvious, since they are not locked up with uncongenial partners. Timmy, a gorilla in the Cleveland Zoo, did not get along with two different female gorillas introduced to him and declined to mate. But when he met a gorilla named Katie, they took to each other at once, playing, mating, and sleeping together. When it was discovered that Katie was infertile, zookeepers decided to send Timmy to another zoo where he might have a chance of breeding and contributing to the gene pool of his endangered species. When a public outcry resulted against the idea of separating Timmy and Katie, the zoo director fumed, "It sickens me when people start to put human emotions in animals, and it demeans the animal. We can't think of them as some kind of mag-nificent human being; they are animals. When people start saying animals have emotions, they cross the bridge of reality." His vehe-ment response shows how strong the fear of anthropomorphism can be even in people who work with animals, notwithstanding their obvious expressions of joy in each other.

Left on their own, are animals faithful to one another? Much has been made of the significant rates of infidelity in some song-birds (both male and female), demonstrated by genetic analysis of parents and offspring and by field observation of the parent birds. Not surprisingly, scientists have not indulged in comic opera ex-periments to see if these songbirds ever resist romantic temptation, though it is known that some animals do seem to reject opportuni-ties for infidelity. For example, male prairie voles who have formed a pair with a female chase away other voles of either sex.

Female African elephants in estrus sometimes form consort-ships with males, but it is not clear whether they simply prefer a certain male and desire only his company, or whether by staying with one male—who vigorously attacks other males that come near—they gain protection from the pursuit of other males. According to Cynthia Moss, consort behavior is something that female ele-phants learn, as young females who do not form a consortship may be "chased and harassed" by many males.

In other species females may not be as vulnerable. Rhinos in the Serengeti, despite their reputation as loners, sometimes do seem to form pair bonds during estrus. One observer saw the male of one pair wander off, whereupon another male appeared and sought to mate, only to be chased off by the female. The male she had paired with then wandered back and they immediately mated. This seems to show that it was not simple lack of interest in mating that caused the female to reject the second male, but a more selective personal principle—perhaps affection—at work. In the light of such observations, it seems possible that among the songbirds whose infidelity has been so carefully studied, those birds who remain faithful to their mates might be exhibiting fidelity based on feelings, rather than lack of other opportunities.

The devotion that members of pairs lavish on one another also gives evidence of love. Some birds are famous examples of fidelity. Geese, swans, and mandarin ducks are all symbols of marital devotion; field biologists tell us this image is accurate. Coyotes, who are considered symbolic of trickery, would make equally good symbols of devotion, since they form lasting pairs. Observations of captive coyotes indicate that they begin to form pair attachments before they are sexually active. Coyote pairs observed by Hope Ryden curled up together, hunted mice together, greeted each other with elaborate displays of wagging and licking, and performed howling duets. Ryden describes two coyotes mating after howling together. Afterward the female tapped the male with her paw and licked his face. Then they curled up together to sleep. This looks a lot like romantic love. Whatever distinctions can be made between the love of two people and the love of two animals, the essence frequently seems the same.

Lovingness may have evolved because animals that had it have been more successful—that is, left more children—than animals that did not. But lovingness is flexible. Instinct may urge the animal to love but does not say who it will love, though it may drop heavy hints. Lovingness instructs an animal to protect and care for its young but does not identify the young. Humans are not so different.

An animal raised by another species will often want a member of that species as its mate when it grows up. A bird, if it is of a species that forms pair bonds, may have the capacity to love a mate. It may also have very specific instincts to court the mate in certain ways. When Tex, a female whooping crane hand-reared by humans, was ready to mate, she rejected male cranes. Instead she was attracted to "Caucasian men of average height with dark hair." Since whooping cranes are so close to extinction, it was considered vital to bring Tex into breeding condition so that she could be artificially inseminated. To do this, International Crane Foundation director George Archibald, a dark-haired Caucasian man, spent many weeks courting Tex. "My duties involved endless hours of 'just being there,' several minutes of dancing early in the morning and again in the evening, long walks in quest of earthworms, nest building, and defending our territory against humans. . . ." The effort was successful and eventually resulted in a crane chick. If the whooping crane dance and nest building are fixed action patterns, the lovingness seems to be a more diffuse impulse. The crane has made an error through no fault of her own. If Tex had been raised by cranes, she would have fallen in love with one, as most cranes do. If George Archibald had been raised by cranes, with whom would he have fallen in love?

Tibby, an otter described in Gavin Maxwell's *Raven, Seek Thy Brother*, was raised by a man who lived on an island off the coast of Scotland, and who got around on crutches. When he became seriously ill, he brought Tibby to Maxwell and asked him to care for her. He died not long thereafter and never returned for her. Tibby did not care for life in the enclosure Maxwell provided, and made a habit of escaping and visiting the nearest village. There she found a man who used crutches and decided to live with him. She tried to build a nest under his house, but he chased her away. A short time later, Tibby disappeared. One day Maxwell received a call from a person who had been alarmed by an otter that had acted strangely, even trying to follow him indoors. Maxwell wrote, "Acting on a sudden inspiration I asked, 'You don't by any chance use crutches, do you?' 'Yes,' he replied, with astonishment in his voice, 'but how in the world could you know that?' " Tibby may have been imprinted on humans who used crutches or she may just have been fond of such people because they reminded her of the affectionate man who had vanished from her life.

While people generally believe other people can and do feel love, we are sometimes skeptical that one particular human loves another, no matter what they say. Parents who do not love their children exist; some children hate their parents, and some husbands and wives or brothers and sisters do not love each other. Yet we continue to believe—to know in our hearts—that there are parents, children, spouses, and siblings who do feel love. It makes sense to use the same standard for animals.

Why has the idea of love among animals been so neglected? Why are we reduced to the rather tedious explanations offered by the evolutionary approach? This is the approach taught in universities, where its wider implications go almost unnoticed. It suggests that the more elaborate the parental tasks an animal performs, the more advantageous it is to have an overarching emotion like love driving them. If parental behavior consists of nothing more than refraining from eating the children, no great emotional drama is necessary. But to feed them, wash them, and risk your life for them —or (perhaps even harder) to let them chew on you, to let them snatch your dinner, and to put up with their noise—you had better love them deeply, at least for the time being. Yet, in the biological analysis, love, however it feels, is principally a device by which subsequent generations are produced. Rather than the reason it exists, this could just be one function it serves. "Scientific" statements about love have captured remarkably little of its essence. Love between two women, between a man and his father, between people and the animals they live with, and love from animal to animal is rarely illuminated by science and more often the subject of wonder and delight in personal statements, poems, novels, and letters. Freeing ourselves from the tyranny of a purely biological explanation might widen the horizon. Love among animals might appear as mysterious and baffling as human love has over the centuries.

Grief, Sadness, and the
Bones of Elephants

In the Rocky Mountains, biologist Marcy Cottrell Houle was observing the eyrie of two peregrine falcons, Arthur and Jenny, as both parents busily fed their five nestlings. One morning only the male falcon visited the nest. Jenny did not appear at all, and Arthur's behavior changed markedly. When he arrived with food, he waited by the eyrie for as much as an hour before flying off to hunt again, something he had never done before. He called out again and again and listened for his mate's answer, or looked into the nest uttering an enquiring "echup." Houle struggled not to interpret his behavior as expectation and disappointment. Jenny did not appear the next day or the next. Late on the third day, perched by the eyrie, Arthur uttered an unfamiliar sound, "a cry like the screeching moan of a wounded animal, the cry of a creature in suffering." The shocked Houle wrote, "The sadness in the outcry was unmistakable; having heard it, I will never doubt that an animal can suffer emotions that we humans think belong to our species alone."

After the cry, Arthur sat motionless on the rock and did not stir for a whole day. On the fifth day after Jenny's disappearance,

Arthur went on a frenzy of hunting, bringing food to his nestlings from dawn to dusk, without pausing to rest. Before Jenny's disappearance, his efforts had been less frenetic; Houle notes that she never again saw a falcon work so incessantly. When biologists climbed to the nest a week after Jenny's disappearance, they found that three of the nestlings had starved to death, but two had survived and were thriving under their father's care. Houle later learned that Jenny had probably been shot. The two surviving nestlings fledged successfully.

It is impossible to predict how deeply affected we will be when somebody close to us dies. Sometimes people show no external reaction, but their lives are shattered. They may feel nothing consciously, or even feel relief, when they are inwardly devastated and may never recover. The external signs of grief tell something, but they may not tell everything. Introspection may tell something, too, but may be misleading. Faced with human sadness in its depths, scientific curiosity should be tempered with humility: no one, certainly not the somatically oriented psychiatrist (offering pills for misery), can speak with any authority about its source, duration, or pathology. Even greater humility is required before the permutations of nonhuman grief and sadness.

When nonscientists speak of animal sadness, the most common evidence they give is the behavior of one of a pair when its mate dies, or the behavior of a pet when its owner dies or leaves. This kind of grief receives notice and respect, yet there are many other griefs that pass unremarked—the cow separated from its calf or the dog deliberately abandoned. Then there are all the griefs humans never see: unheard cries in the forest, herds in the remote hills whose losses are unknown.

Mourning Lost Love

Wild animals have been observed mourning for a mate. According to naturalist Georg Steller, the now-extinct sea cow named after him was a monogamous species, with families consisting usually of a female, a male, and two young of different ages: "one grown offspring, and a little, tender one." Steller, a ship's naturalist, saw that when the crew of the ship killed a female whose body washed up on the beach, the male returned to her body for two consecutive days, "as if he were inquiring about her."

As the fate of the three dead peregrine nestlings shows, it can be disastrous for a wild animal to display grief. There is no survival value in not eating, or moping and grieving. While love can be readily reduced to evolutionary function (for those so inclined), grief over the loss of a loved one—another expression of love— often threatens survival. Grief thus calls for explanation on its own terms.

The sorrow of bereavement is easily observed in captive or pet animals. Elizabeth Marshall Thomas gives a moving account of Maria and Misha, two huskies who had formed a pair bond, when Misha's owners gave him away.

> Both he and Maria knew that something was terribly wrong when his owners came for him the last time, so that Maria struggled to follow him out the door. When she was prevented, she rushed to the window seat and, with her back to the room, watched Misha get into the car. She stayed in the window for weeks thereafter, sitting backward on the seat with her face to the window and her tail to the room, watching and waiting for Misha. At last she must have realized that he wasn't going to come. Something happened to her at that point. She lost her radiance and became depressed. She moved more slowly, was less responsive, and got angry rather easily at things that before she would have overlooked. . . . Maria never recovered from her loss, and although she never forfeited her place as alpha female, she showed no interest in forming a permanent bond with another male. . . .

Maria knew that Misha was gone from her. Her behavior is reminiscent of human grief at permanent separation and loss of a loved one. Wolves and coyotes, to whom dogs are very closely related, do form pairs. The conditions in which dogs are kept are very different from those in which wild canids live. Probably dog behavior is more flexible than has been realized, more strongly dictated by the conditions humans provide for them. While both female and male dogs have come to symbolize promiscuity to many humans, this behavior has been created by the way humans breed and maintain dogs; it is not intrinsic to their nature. One has to wonder how much so-called "natural" human sexuality is equally produced by social arrangements and expectations.

Some animals who do not form pairs in the wild are housed in pairs in captivity and grow deeply attached to one another. Often the mate is the only companion the animal has. Ackman and Alle, two circus horses, were stabled together. No particular attachment between them was noticed until Ackman's unexpected death. Alle "whinnied continually." She scarcely ate or slept. In an effort to distract her, she was moved, given new companions, and offered special foods. She was examined and medicated, in case she was ill. Within two months she had wasted to death.

Two Pacific "kiko" dolphins in a marine park in Hawaii, Kiko and Hoku, were devoted to each other for years, often making a point to touch one another with a fin while swimming around in their tank. When Kiko suddenly died, Hoku refused to eat. he swam slowly in circles, with his eyes clenched shut "as if he did not want to look on a world that did not contain Kiko," as trainer Karen Pryor wrote. He was given a new companion, Kolohi, who swam beside him and caressed him. Eventually he opened his eyes and ate once more. Although he became attached to Kolohi, observers felt that he never became as fond of her as he had been of Kiko. While the interpretation that Hoku did not want to see a world without Kiko remains speculative, it is clear that Hoku was grieving.

Researchers who had caught a dolphin on a fishhook and put her in a holding tank soon despaired for her life. Pauline, as they named her, could not even keep herself upright and had to be supported constantly. On the third day of her captivity, a male dolphin was captured and placed in the same tank. This raised her spirits; the male helped her swim, at times nudging her to the surface. Pauline appeared to make a complete recovery but suddenly died two months later from an abscess caused by the fishhook, whereupon the male refused to eat and died three days later. An autopsy revealed a perforated gastric ulcer, surely aggravated by his mournful fasting.

It would be the end of most species if every bereaved animal died of grief. Such cases must be extreme and unusual. Dying of grief is not the only proof of love and affection in animals, but these incidents do illuminate an emotional range and emotional possibilities. Animals in the wild also grieve for companions other than mates. Lions do not form pairs, yet a lion has been known to remain by the body of another lion that had been shot and killed, licking its fur. As is so often the case, elephants offer examples that are uncannily similar to human feelings. Cynthia Moss, a researcher who has studied wild African elephants for years, describes mother elephants who appear in perfect health but become lethargic for many days after a calf dies and trail behind the rest of the family.

An observer once came across a band of African elephants surrounding a dying matriarch as she swayed and fell. The other elephants clustered around her and tried mightily to get her up. A young male tried to raise her with his tusks, put food into her mouth, and even tried sexually mounting her, all in vain. The other elephants stroked her with their trunks; one calf knelt and tried to suckle. At last the group moved off, but one female and her calf stayed behind. The female stood with her back to the dead matriarch, now and then reaching back to touch her with one foot. The other elephants called to her. Finally, she walked slowly away.

Cynthia Moss describes the behavior of an elephant herd circling a dead companion "disconsolately several times, and if it is still motionless they come to an uncertain halt. They then face outward, their trunks hanging limply down to the ground. After a while they may prod and circle again, and then again stand, facing outward." Finally—perhaps when it is clear the elephant is dead— "they may tear out branches and grass clumps from the surrounding vegetation and drop these on and around the carcass." The standing outward suggests that the elephants may find the sight painful; maybe they want to stay close but find it intrusive to watch such suffering; perhaps it has a ritual meaning we do not yet comprehend.

It was once thought that elephants went to special elephant graveyards to die. While this has been disproved, Moss speculates that elephants do have a concept of death. They are strongly interested in elephant bones, not at all in the bones of other species. Their reaction to elephant bones is so predictable that cinematographers have no difficulty filming elephants examining bones. Smelling them, turning them over, running their trunks over the bones, the elephants pick them up, feel them, and sometimes carry them off for a distance before dropping them. They show the greatest interest in skulls and tusks. Moss speculates that they are trying to recognize the individual.

Once Moss brought the jawbone of a dead elephant—an adult female—into her camp to determine its exact age. A few weeks after this elephant's death, her family happened to pass through the camp area. They made a detour to be with and examine the jaw. Long after the others had moved on, the elephant's seven-year-old calf stayed behind, touching the jaw and turning it over with his feet and trunk. One can only agree with Moss's conclusion that the calf was somehow reminded of his mother—perhaps remembering the contours of her face. He felt her there. It seems certain that the calf's memory is at work here. Whether he experienced a feeling of melancholic nostalgia, sorrow, perhaps joy in remembering his mother, or was moved by some emotional experience we might not be able to identify, it would be difficult to deny that feelings were involved.

Based on their behavior, the feelings of the Gombe chimpanzees who witnessed one of their number fall to his death seem similarly complex. Three small groups of chimpanzees, mostly males, but including a female in estrus, had come together when Rix, an adult male, somehow fell into a rocky gully and broke his neck. The reaction was immediate pandemonium—apes screaming, charging, displaying, embracing, copulating, throwing stones, barking, and whimpering seemingly at random. Eventually they grew quieter. For several hours the chimps gathered around the corpse. They came close and peered at Rix's body silently, climbed on branches to get a different vantage point. They never touched him. One male adolescent, Godi, seemed particularly intent, whimpering and groaning repeatedly as he gazed at Rix. Godi became highly agitated when several large males drew very close to the body. After several hours the chimpanzees drifted away. Before leaving, Godi leaned over Rix and stared at him intently before hurrying after his companions.

During this episode the chimpanzees repeatedly uttered "wraah" calls, common when chimps are disturbed by strange humans or by Cape buffalo, but also when they sight a dead chimp or baboon. On one occasion chimps *wraah*ed for four hours after witnessing the death of a baboon injured in a fight with other baboons.

A chimpanzee at the Arnhem Zoo, rather confusingly named Gorilla, had several babies who died despite her tender care. Each time an infant died she would become visibly depressed. Gorilla would sit huddled in a corner for weeks on end, ignoring the other chimpanzees. At times she would burst out screaming. This story had a happy sequel: Gorilla became a successful mother when she was given the care of Roosje, a ten-week-old baby chimpanzee, whom she was taught to bottle-feed.

Loneliness

Loneliness appears to affect animals who live in social or family groups. It is probably one factor causing death in many captive animals. For captive beavers, for example, the presence or absence of a companion is an important factor in survival. One wildlife biologist noted that yearling beavers, "if they do not get companionship, may simply sit where they are put down until they die." Loneliness, a frequent result of confinement and domesticity, is often observed in captive animals. A lonely wild beaver could presumably set off in search of other beavers.

Animals seek each other out more than biologists once assumed, perhaps in an effort to avoid feelings of sadness, loneliness, and sorrow. In some species, males who have been "kicked out of the nest" by their mothers form bachelor herds. Male African elephants gather in groups in "bull areas." Many animals are rather sweepingly described as solitary, but careful field studies of animals famous for their solitary natures—tigers, leopards, rhinoceros, and bears—often reveal that they spend more time associating with each other than was previously thought. The European wildcat and the fishing cat are said to be solitary species in which the female and male mate and then separate, and the female raises the kittens alone. In zoos, however, the female and male may be caged in pairs, with interesting results. Usually the male is taken out before kittens are born, in case he should harm them. In the Cracow Zoo this precaution was omitted, and instead of attacking the kittens the male wildcat carried his meat to the entrance of the den and made coaxing sounds. Similarly, in the Magdeburg Zoo the father wildcat guarded the den day and night and, though normally peaceful, attacked the keeper if he came too near. The father brought food to the den, and when the kittens were old enough to come out and play, he hissed and threatened any zoo-goers who startled his kittens. Fishing cats at the Frankfurt Zoo also led a surprisingly warm family life. The male not only brought food, but often curled up in the nest box with the rest of the family. He was such a conscientious parent that if he was out of the nest box and the female also came out, he became anxious and went in the nest box with the kittens.

Possibly these species are less solitary than has been thought, or perhaps this is another demonstration of the flexibility of animal behavior. Paul Leyhausen, who observed these cats, speculates that while males in the wild may have nothing to do with their mates and kittens, in captivity they may be "subjected to stimuli which awaken normally dormant behavior patterns." If so, we are entitled to wonder whether a male fishing cat, wandering by a southeast Asian stream or through a forest, ever gets a twinge from those normally dormant patterns, and feels lonely.

Imprisonment

Even when captive animals are not confined in solitude, their imprisonment may make them sad. It is often said of zoo animals that the way to tell if they are happy is to ask whether the young play and the adults breed. Most zookeepers would not accept this standard of happiness for themselves. As Jane Goodall noted, "Even in concentration camps, babies were born, and there is no good reason to believe that it is different for chimpanzees."

Captivity is undoubtedly more painful to some animals than others. Lions seem to have less difficulty with the notion of lying in the sun all day than do tigers, for example. Yet even lions can be seen in many zoos pacing restlessly back and forth in the stereotyped motions seen in so many captive animals. The concept of *funktionslust*, the enjoyment of one's abilities, also suggests its opposite, the feeling of frustration and misery that overtakes an animal when its capacities cannot be expressed. If an animal enjoys using its natural abilities, it is also possible that the animal *misses* using them. Although a gradual trend in zoo construction and design is to make the cages better resemble the natural habitat, most zoo animals, particularly the large ones, have little or no opportunity to use their abilities. Eagles have no room to fly, cheetahs have no room to run, goats have but a single boulder to climb.

There is no reason to suppose that zoo life is not a source of sadness to most animals imprisoned there, like displaced persons in wartime. It would be comforting to believe that they are happy there, delighted to receive medical care and grateful to be sure of their next meal. Unfortunately, in the main, there is no evidence to suppose that they are. Most take every possible opportunity to escape. Most will not breed. Probably they want to go home. Some captive animals die of grief when taken from the wild. Sometimes these deaths appear to be from disease, perhaps because an animal under great stress becomes vulnerable to illness. Others are quite obviously deaths from despair—near-suicides. Wild animals may refuse to eat, killing themselves in the only way open to them. We do not know if they are aware that they will die if they do not eat, but it is clear that they are extremely unhappy. In 1913 Jasper Von

Oertzen described the death of a young gorilla imported to Europe: "Hum-Hum had lost all joy in living. She succeeded in living to reach Hamburg, and from there, the Animal Park at Stellingen, with all her caretakers, but her energy did not return again. With signs of the greatest sadness of soul Hum-Hum mourned over the happy past. One could find no fatal illness; it was as always with these costly animals: 'She died of a broken heart.' "

Marine mammals have a high death rate in captivity, a fact not always apparent to visitors at marine parks and oceanariums. A pilot whale celebrity at one oceanarium was actually thirteen different pilot whales, each successive one being introduced to visitors by the same name, as if it were the same animal. It takes little reflection to see the great difference in a marine mammal's life when kept in an oceanarium. Orcas grow to twenty-three feet long, weigh up to 9,000 pounds, and roam a hundred miles a day. No cage, and certainly not the swimming pools where they are confined in all oceanariums, could possibly provide satisfaction, let alone joy. They are believed to have a life expectancy as long as our own. Yet at Sea World, in San Diego, the oceanarium with the best track record for keeping orcas alive, they last an average of eleven years.

If a person's life span were shortened this much, would one still speak of happiness? Asked whether their animals were happy, a number of marine mammal trainers all said yes: they ate, engaged in sexual intercourse (it is extremely rare for an orca to give birth in captivity), and were almost never sick. This could mean that they were not depressed, but does it mean they were happy? The fact that people ask this question again and again indicates a malaise, perhaps profound guilt at subjecting these lively sea travelers to unnatural confinement.

Which animals suffer the most in captivity can be unpredictable. Harbor seals often thrive in oceanariums and zoos. Hawaiian monk seals almost invariably die—sometimes they refuse to eat, sometimes they succumb to illness. One way or another, one observer noted, they have generally "just moped to death."

The issue of the effects of captivity is most painful when one considers animals that can live nowhere but in captivity because their habitat is gone—as is the case for an increasing number of species—or because they are physically incapacitated. When fewer than a dozen California condors were left in the wild, arguments raged about whether to capture the remaining birds for captive breeding or to let the species perish freely, without undergoing the ignominy of captivity. The condor is a soaring bird that can easily fly fifty miles in a day, a life that can hardly be simulated in a cage. In the end the birds were captured, and so for a time there were no California condors in the wild. Since then, captive-bred birds have been released in an attempt to reestablish the species.

The fact that animals *can* be sad must first be acknowledged before it can be studied and understood. Zookeepers ask whether animals are healthy, and whether they are likely to breed, but rarely ask, "What would make this animal *happy?*" Nor have the studies by animal behaviorists been of much help. The *Oxford Dictionary of Animal Behavior* notes: "It seems reasonable to allow that animals may be distressed by being unable to feed and drink, to move their limbs, to sleep, and to have social interaction with their fellows, but the difficulty of defining distress in an objective and convincing way has been a stumbling-block in the formation of animal welfare legislation even in countries where there is widespread public interest in the way that animals are treated."

Depression and Learned Helplessness

In humans, extreme sadness is called depression. As used by psychiatrists and psychologists, depression is a catchall diagnosis, referring to melancholy springing from a number of sources. In the quest to validate the medical model of psychiatry, scientists have sought to produce clinically depressed animals in the laboratory— to which end some experimenters have worked to provide animals with spectacularly unhappy childhoods.

Among the most widely reported experiments in the history of animal behavior are those psychologist Harry Harlow performed on rhesus monkeys. The baby monkeys under his aegis who preferred soft, huggable dummy mothers to hard, wire surrogates, even when only the wire ones dispensed milk, are famous and have been used as evidence that psychological studies on animals—really forms of torture—can teach humans about their emotions. While the study suggests that the nurturant feelings in mothering can be even more important than its survival value, surely this gruesome experiment was gratuitously emotionally cruel as well as unnecessary to prove this point.

Other rhesus monkeys, at the age of six weeks, were placed alone in the "depression chamber," or vertical chamber, a stainless-steel trough intended to reproduce a psychological "well of despair." Forty-five days of solitary confinement in the chamber produced permanently impaired monkeys. Even when months had passed since their experience, the once-chambered monkeys were listless, incurious, and almost completely asocial, huddling in one spot and clasping themselves. No knowledge gained, no point proved, can justify such abuse.

Similarly, dogs, cats, and rats in the laboratory have been induced to feel the global pessimism known as "learned helplessness." In the classic experiment, dogs were strapped into a harness and given electric shocks at unpredictable intervals. The shock was inescapable—nothing they could do prevented or lessened it. Afterward they were placed in a divided chamber. When a tone sounded, the dogs needed to jump into the other side of the chamber to avoid being shocked. Most dogs learned this quickly, but two thirds of the dogs who had been given inescapable shocks just lay still and whined, making no attempt to escape. Their previous experience had apparently taught them despair. This effect wore off in a few days. Yet, if the dogs were subjected to inescapable shocks four times in a week, their "learned helplessness" was lasting. Psychologist Martin Seligman, the principal researcher in the study of learned helplessness (and author of the bestselling book *Learned Optimism*), argues that the shocked animal is frightened at first, but when it comes to believe that it is helpless, sinks into depression. In his explanation of how he came upon the notion of doing experiments on learned helplessness in animals Seligman cites the research of C. P. Richter during the fifties, "who reasoned that for a wild rat, being held in the hand of a predator like man, having whiskers trimmed, and being put in a vat of hot water from which escape is impossible produces a sense of helplessness in the rat."

Learned helplessness has been experimentally produced in humans, though not by means of shock. People given tasks at which they repeatedly fail quickly come to believe that they will fail at other tasks and do poorly at them, compared to those who have not been put through a sequence of failures. In the real world, battered women may be unable to leave their batterers, although the risks of leaving and the lack of anywhere to go may be as important as their perception that any action on their part to save themselves from continued abuse is pointless. The animal research really shows nothing about humans—the alleged purpose of such research—that one cannot learn by talking to battered women about their lives.

Having produced depressed dogs, Seligman wanted to cure them. He placed "helpless" dogs in the chamber and removed the partition to make it easy for them to cross and avoid shock, but the despondent dogs made no effort to get away and so did not discover escape was possible. Seligman got in the chamber and called them, and offered them food, but the dogs did not move. Eventually he was reduced to dragging the dogs back and forth on leashes. Some dogs were dragged back and forth two hundred times before they discovered that this time they could escape the electric shocks. According to Seligman, their recovery from learned helplessness was lasting and complete. Their experience, however, must have had some lasting effect on them.

Many other experimenters have produced learned helplessness in the laboratory by various means, with sometimes fiendish results. One experimenter raised rhesus monkeys in solitude, in black-walled isolation cages, from infancy until six months, to induce "social helplessness." Then he taped each young monkey to a cruciform restraining device and placed it, for an hour a day, in a cage with other young monkeys. After initial withdrawal, the unrestrained monkeys poked and prodded the restrained monkeys, pulling their hair, gouged their eyes, and pried their mouths open. The restrained monkeys struggled, but could not escape. All they could do was cry out. After two to three months of this abuse, their behavior changed. They stopped struggling, though they still cried out. And as the experimenter noted, "No advantage was taken of numerous opportunities to bite the oppressor which thrust fingers or sex organs against or into its mouth." These monkeys were lastingly traumatized and were terrified of other monkeys even when unrestrained. Like the other experiments, this one is distinguished by its cruelty.

Comparatively few depressed humans became so through being placed in solitary confinement for half their childhoods, or by being raised in solitary confinement and then tortured by peers. Oddly enough, the argument on the part of the scientists conducting these experiments has been that animals are so similar to us in their feelings that we can learn about human depression by studying animal depression. But this raises the important ethical question asked by many animal-rights groups: If animals suffer the way we do, which is the whole justification for the experiments, is it not sadistic to conduct them? Clearly the animals can be made deeply unhappy, but this fact could have been observed under naturally occurring conditions, without subjecting sensitive creatures to pointless cruelty.

Through all these griefs and torments, animals display sorrow through their movements, postures, and actions. Often animal vocalizations provide evidence of sadness. Wolves seem to have a special mourning howl or lonesome howl that differs from their usual convivial howling. Other animals are said to wail, moan, or cry. When Marchessa, an elderly female mountain gorilla, died, the silverback male of her group became subdued and was heard to whimper frequently, the only time such a sound had been heard from a silverback. These two wild gorillas may have spent as much as thirty years of their lives together. One observer wrote of orangutans, "In disappointment the young specimen quite commonly whimpers or weeps, without, however, shedding tears."

No one knows for certain why humans weep. Newborn babies cry, but they do not usually shed tears until they are a few months old. Adults cry less, and some adults never shed tears. Tears have been classified into three kinds: continuous tears, which keep the eye moist; reflex tears, which flush foreign objects or irritating gases out of the eye; and emotional tears, the tears of grief, happiness, or rage. Emotional tears are different in that they contain a higher percentage of protein than other tears. Curiously, since Darwin's survey of the subject in 1872, weeping has been little studied, but it has been speculated that emotional tears may have both physical and social or communicative functions.

Since it is possible for people to feel great unhappiness and not weep, it is also unclear why tears communicate so effectively. It may be that our reaction is instinctive, and perhaps part of the respect accorded to tears comes from the possibility that they are ours alone. It has been suggested that almost every human bodily secretion is considered disgusting (such as feces, urine, and mucus), and its ingestion is taboo with one exception: tears. This is the one body product that may be uniquely human and hence does not remind us of what we have in common with animals.

Perhaps it isn't only humans who are impressed by tears, however. The chimpanzee Nim Chimpsky, who regularly sought to comfort people who looked sad, was particularly tender when he saw tears, which he would wipe away. Since Nim was raised by humans, he may have learned the connection between tears and unhappiness.

It would be interesting to discover whether any animals who have not had the opportunity to learn about tears respond to tears as evidence of sadness in humans or even in other animals. This could be answered experimentally. If a chimpanzee reared with other chimps saw another who appeared to be shedding tears, would it react as Nim did? If a chimpanzee accustomed to humans saw a person cry for the first time, would it behave as though it was a sign of distress?

Tears keep the eyes of animals moist. Their eyes also water when irritated. Tears may spill from the eyes of an animal in pain. Tears have been seen in the eyes of animals as diverse as an injured horse and an eggbound grey parrot. Some animals are more tearful than others. Seals, who have no nasolachrymal ducts into which tears drain, are especially apt to have tears rolling down their faces. This is thought to help them cool down when they are on land.

Charles Darwin, in researching *The Expression of the Emotions in Man and Animals,* looked for evidence that animals did or did not shed emotional tears. He complained, "The *Macacus maurus,* which formerly wept so copiously in the Zoological Gardens, would have been a fine case for observation; but the two monkeys now there, and which are believed to be of the same species, do not weep." He was not able to observe animals shedding emotional tears, and called weeping one of the "special expressions of man."

Darwin noted one exception: the Indian elephant. It was reported to him by Sir E. Tennant that some newly captured elephants in Ceylon (now Sri Lanka), tied up and lying motionless on the ground, showed "no other indication of suffering than the tears which suffused their eyes and flowed incessantly." Another captured elephant, when bound, sank to the ground, "uttering choking cries, with tears trickling down his cheeks." A captured elephant is usually also separated from its family. Other elephant observers in Ceylon assured Darwin that they had not seen elephants weep, and that Ceylonese hunters said they had never seen elephants weep. Darwin put his trust in Tennant's observations, however, because they were confirmed by the elephant keeper at the London Zoo, who said he had several times seen an old female there shedding tears when her young companion was taken out.

In the years since Darwin, the balance of the evidence has been the same: most elephant watchers have never seen them weep —or have, rarely, seen them weep when injured—yet a few observers have claimed to have seen them weep when not injured. An elephant trainer with a small American circus told researcher William Frey that his elephant, Okha, does cry at times, but that he had no idea why. Okha sometimes shed a tear when being scolded, it is reported, and at least once wept while giving children rides. Iain Douglas-Hamilton, who has spent years working with African elephants, has seen elephants shed tears only when injured. Tears fell from the eyes of Claudia, a captive elephant, during a difficult labor with her first calf.

R. Gordon Cummings, an eighteenth-century hunter in South Africa, described killing the biggest male elephant he had ever seen. He first shot it in the shoulder so that it could not run away. The elephant limped over to a tree and leaned against it. Deciding to contemplate the elephant before killing it, Cummings paused to make coffee and then chose to experimentally determine which were an elephant's vulnerable spots. He walked up to it and fired bullets into various parts of the head. The elephant did not move except to touch the bullet wounds with the tip of his trunk. "Surprised and shocked to find that I was only tormenting and prolonging the sufferings of the noble beast, which bore his trials with such dignified composure," Cummings wrote, he decided to finish him off and shot him nine times behind the shoulder. "Large tears now trickled from his eyes, which he slowly shut and opened; his colossal frame quivered convulsively, and, falling on his side, he expired." This elephant must have been in great pain, however, and that alone would have been cause enough for him to shed tears. Other than humans, no animal runs torture experiments on other animals.

In his book *Elephant Tramp*, George Lewis, an itinerant elephant trainer, reported in 1955 that in the years he had worked with elephants he had seen only one weeping. This was a young, timid female named Sadie, who was being trained along with five others to do an act for the Robbins Brothers Circus. The elephants were being taught their acts quickly, since the show would start in three weeks, but Sadie had trouble learning what was wanted. One day, unable to understand what she was being told to do, she ran out of the ring. "We brought her back and began to punish her for being so stupid." (Based on information Lewis gives elsewhere, they probably punished her by hitting her on the side of the head with a large stick.) To their astonishment, Sadie, who was lying down, began to utter racking sobs, and tears poured from her eyes. The dumbfounded trainers knelt by Sadie, caressing her. Lewis says that he never punished her again, and that she learned the act and became a "good" circus elephant. His fellow elephant trainers, who had never witnessed such a thing, were skeptical. But reports are not confined to animal behaviorists. Victor Hugo wrote in his diary on January 2, 1871: "*On a abbabut l'éléphant du Jardin des*

Plantes. Il a pleuré. On va le manger." ["The elephant in the Jardin des Plantes was slaughtered. He wept. He will be eaten."] That elephants weep emotional tears is widely believed in India, where elephants have been kept for many centuries. It is said that when the conqueror Tamerlane captured three thousand elephants in battle, snuff was put in their eyes so they would appear to be weeping at the loss. Douglas Chadwick was told of a young Indian elephant shedding tears when scolded for playing too boisterously and knocking someone down, and also of an elephant that ran away and, when found by its mahout, wept along with him. Observing young orphaned elephants in an Asian stable, Chadwick noticed that one was shedding tears. A mahout told him that the babies often cried when they were hungry and that it was almost feeding time. But after being fed, the baby still wept.

Elephant handlers say that the eyes of elephants water heavily, presumably to keep them moist. Fluid may also stream from their temporal glands, which are between the eye and ear. But no one familiar with elephants would be confused by this. Possibly there is some significance to the fact that many of the elephants shedding tears were lying down, not a usual position for an elephant. Perhaps the position somehow prevents drainage of tears. For all we know elephants often shed tears of grief, but if standing, the tears run through nasolachrymal ducts and down the inside of their trunks.

Emotional tears have been reported in some other species. Biochemist William Frey, who studies human emotional tears, has received reports of dogs—particularly poodles—shedding tears in emotional situations, such as being left behind by their owner, but despite repeated efforts, he has been unable to confirm this in the laboratory. No one but their owners has witnessed these tears, and poodles are a particularly damp-eyed breed even at their most cheerful.

It has been reported that tears rolled from the eyes of adult seals who saw seal pups clubbed by hunters. This is undoubtedly true. But since tears often roll from seals' eyes, there is no proof that these were emotional tears.

Beavers have also been suspected of crying emotional tears. Trappers have said that a beaver in a trap sheds tears, but such beavers may be crying in pain. However, one biologist has reported that beavers also weep copiously when manually restrained. Dian Fossey reported tears shed by Coco, an orphaned mountain gorilla. Coco was three or four years old when her family was killed before her eyes to secure her capture. She had spent a month in a tiny cage before coming into Fossey's possession and was very ill. She was released into an indoor pen with windows. When Coco first looked out the window of her pen at a forested mountainside like the one on which she grew up, she suddenly began "to sob and shed actual tears." Fossey said she never witnessed a gorilla do this before or afterward.

Montaigne, who may be the first Western author to express distaste for the hunt, wrote in his 1580 essay "Of Cruelty":

> For myself, I have not even been able without distress to see pursued and killed an innocent animal which is defenseless and which does us no harm. And as it commonly happens that the stag, feeling himself out of breath and strength, having no other remedy left, throws himself back and surrenders to ourselves who are pursuing him, asking for our mercy by his tears . . . that has always seemed to me a very unpleasant spectacle.

In the end, it hardly matters whether stags, beavers, seals, or elephants weep. Tears are not grief, but tokens of grief. The evidence of grief from other animal behaviors is strong. It is hard to doubt that Darwin's sobbing elephants were unhappy, even if their tears sprang from mechanical causes. A seal surely feels sad when its pup is killed, whether it is dry-eyed or not. Just as a psychiatrist cannot really know when a person has crossed the border of "normal" grief to "pathological" mourning, so humans cannot know that the world of sorrow is beyond the emotional capacities of any animal. Sadness, nostalgia, disappointment, are feelings we know from direct experience; animals we know intimately hint at their parallel feelings in this dark world. Should science accept their challenge and try to understand animal sorrow, even its accurate description will need to be complex and subtle, well beyond the clumsy categories and reductive causalities that prevail in the psychology of human pain.

A Capacity for Joy

Far out at sea, a tuna fleet surrounded a group of spinner dolphins swimming over a school of tuna, catching them in a gigantic net. Small, powerful speedboats circled the animals, creating a wall of sound that disoriented and terrified the dolphins, who sank down silently into the net, only the movement of their eyes showing signs of life. Biologists trying to learn how to save dolphins looked on in despair. But when a dolphin crossed the corkline at the edge of the net, "It *knew* it was free. It burst forward, propelled by powerful wide-amplitude tail strokes . . . [it] then dove, swimming at full speed . . . down and away into the dark water, only to burst from the surface in a high bounding series of leaps."

In an account of this episode, dolphin biologist Kenneth Norris focused on the state of the trapped dolphins, persuasively arguing that their behavior demonstrated not apathy but deep fear. Equally compelling is the joy of the freed dolphins, springing through air and water.

Theorists of human joy have sought to categorize it and to analyze its causes in terms that range from a "sharp reduction in the gradient of neural stimulation" to "what obtains after some

_____ 111

creative or socially beneficial act that was not done for the express purpose of obtaining joy or doing good." Such theorists tend to ignore the possibility that animals, too, feel joyous.

No one who has ever had a dog or cat can doubt the animal capacity for happiness. Beholding and sharing their open joy is one of the great pleasures people take in animals. We see them leap or run, hear them bark or chirrup, and put words to their delight: "You're home!" "You're going to feed me!" "We're going for a walk!" Like uninhibited human happiness, the pleasure is contagious, so that pets serve as a conduit to joyful feelings. It is rare to find a person as openly ecstatic as a cat about to be fed, or a dog about to go for a walk. If such joy were a figment of anthropomorphic projection, it would be a remarkable collective delusion.

Happiness can be a reward, a response of pleasure in accomplishment. If an animal feels good from doing things that have selective value, certainly that happiness can be said to have selective value. But that does not necessarily mean that the happiness exists only because it has selective value. The grim tasks of survival, even surviving well, do not make a lot of people happy. Part of happiness is often its lack of relation, or even its perverse relation, to any rational end, its utter functionlessness. The evidence is good that animals as well as people do feel such pure joy.

One of the many signs by which joy in animals can be recognized is vocalization. Pet cats are admired for purring, a sound that usually indicates contentment, though it may also be used to appease another animal. Big cats purr too. Cheetahs purr loudly when they lick each other, and cubs purr when they rest with their mother. Lions purr, though not as often as house cats, and only while exhaling. Both young and adult lions also have a soft hum they utter in similar circumstances—when playing gently, rubbing their cheeks together, licking each other, or resting.

Happy gorillas are said to sing. Biologist Ian Redmond reports that they make a sound—something between a dog whining and a human singing—when they are especially happy. On a rare sunny day, when the foraging is particularly good, the family group will eat, "sing" and put their arms around each other. Howling wolves may be asserting territorial rights or cementing social bonds, but observers say it also appears to make them happy.

Black bear cubs express their emotions more clearly than adults, says wildlife biologist Lynn Rogers. "When a cub is very comfortable, particularly when it's nursing, they give what I call a comfort sound. I used to call it a nursing vocalization until I saw cubs doing it when they were not nursing." He imitates the sound, a low squeal. "A pleasing little sound that they make. One time I gave a big bear a piece of warm fat. It just really seemed to like it. In fact, it made that same sound in a deeper voice. So I don't know —is that happiness or not? Is it just comfort? It was pleased, anyway."

Joy can also be expressed silently. Observers of almost any species will quickly learn to know the body language of a happy animal. Darwin cited the frisking of a horse turned out to pasture and the grins of orangutans and monkeys being caressed. In a personal letter he also gave a charming account of animal joy:

> Two days since, when it was very warm, I rode to the Zoological Society, & by the greatest piece of good fortune it was the first time this year, that the Rhinoceros was turned out.—Such a sight has seldom been seen, as to behold the rhinoceros kicking & rearing, (though neither end reached any great height) out of joy.—The elephant was in the adjoining yard & was greatly amazed at seeing the rhinoceros so frisky: He came close to the palings & after looking very intently, set off trotting himself, with his tail sticking out at one end & his trunk at the other,—squeeling and braying like half a dozen broken trumpets.

Signs of happiness can no doubt be misinterpreted. One of the many factors contributing to the fascination with bottle-nosed dolphins is their permanent "smile," created by the shape of their jaws rather than by an emotional state. Since a dolphin doesn't have a mobile face, it "smiles" even if furious or despondent.

Despite this, biologist Kenneth Norris believes that people and dolphins can recognize the emotional freight of many of each other's signals. That is, the two species can recognize or learn to recognize friendliness, hostility, or fear across species, even if we do not understand each other's vocalizations. He cites a spinner dolphin's "peremptory" barks, which indicate boisterous behavior, as compared with soft chuckles that indicate friendly contact, often between female and male. The body language of human and dolphin mothers with their babies, Norris says, is not only comparable, but easily understood by both species.

When the ice finally melted from one New England beaver pond in spring, a male beaver and his yearling daughter swam over to look at their dam, "porpoising" on the way, swimming over one another's backs. Afterward they swam across the pond together, rolling, diving, popping up again and turning somersaults, in a display of delight with which even a nonswimmer could identify.

In an example of more literal language, several apes have been taught the sign for "happy." Nim Chimpsky used the word when he was excited, as when he was being tickled. Koko, asked what gorillas say when they're happy, signed "gorilla hug." Whether Nim and Koko would understand one another's use of "happy" is unknown. One criticism of ape-language teaching has been that the animals, with the exception of the gorilla Koko, were usually not taught words to express emotion, though it seemed likely that they would want to communicate emotive states to their friends and enemies. Carolyn Ristau suggested: "It could be worthwhile to attempt teaching a chimpanzee to associate signs with such mental states as aggressive, frightened, in pain, hungry, thirsty or wanting to play." These might be the words that would interest them deeply.

One rainy day in Washington the signing chimpanzees Moja and Tatu were offered a chance to go outside into an exercise area. Moja, who hates rain, went out but scurried into a cave. Tatu climbed to the top of a play structure and sat in the rain, signing "out out out out out out." A researcher said, "It looked like she was singing in the rain."

An equally expressive behavior called "war dancing" is seen in mountain goats and chamois. One animal starts rearing, leaping, tossing its horns, and whirling about. One after another, the whole band takes it up. Goats war dance most often in the summer, when food is plentiful. The sight of a slanting snow bank can start a war dance and send the band bucking, twirling, and sliding downhill, kicking up the snow. They expend so much energy in their dance that some goats have been seen to make almost two full turns in the air in one jump.

What have these goats got to be happy about? They have not heard news of an inheritance, received a job offer, or seen their names in the newspaper. They have nothing to be pleased about except life, sunshine, and being well fed. They jump for joy.

Sometimes the source of the joy is obvious and recognizable, such as the excitement displayed by a group of wild chimpanzees finding a large pile of food. "Three or four adults may pat each other, embrace, hold hands, press their mouths against one another, and utter loud screams for several minutes before calming down sufficiently to start feeding," Goodall and Hamburg reported. The implications seemed obvious. "This kind of behavior," they wrote, "is similar to that shown by a human child, who, when told of a special treat, may fling his arms ecstatically around the bearer of the good news and squeal with delight."

A principal source of joy for social animals is the presence of their family and the members of their group. Nim Chimpsky was raised in a human family for the first year and a half of his life. When he was about four years old, a reunion was arranged with the family that had raised him. When he spotted them, in a place where he had never seen them before, Nim smiled hugely, shrieked, and pounded the ground for three minutes, gazing back and forth at the different members of the family. Finally he calmed down enough to go and hug his foster mother, still smiling, and shrieking intermittently. He spent more than an hour hugging his family, grooming them and playing with them before they left. This was the only occasion on which Nim was seen to smile for more than a few minutes.

Reunions after separation are a common source of joy. Two male bottle-nosed dolphins at an oceanarium did not have the adversarial relationship seen between many male dolphins confined together. One was removed to another exhibit for three weeks.

When he was returned, the two seemed very excited. For hours they hurtled around the tank side by side, occasionally leaping out of the water. For several days they spent all their time together, ignoring the other dolphin in the tank.

The meeting of two related groups of elephants seems to be a very emotional time, full of ecstasy and drama. Cynthia Moss has reported the meeting of two such groups, one led by the old female Teresia, the other led by Slit Ear. From a quarter of a mile away they began calling to each other. (Since elephants can communicate over long distances with sounds too low for us to hear, they might have been aware of each other before they started calling audibly.) Teresia changed direction and began walking fast. Their heads and ears were up, and fluid poured from the temporal glands (small glands between the eye and ear) of all the elephants in the herd. They stopped, called, got a response, changed course slightly, and sped ahead. Slit Ear's group appeared out of some trees, running toward them.

The groups ran toward each other, screaming and trumpeting. Teresia and Slit Ear rushed together, clicked tusks and twined trunks together while rumbling and flapping their ears. All the elephants performed similar greetings, spinning around, leaning on each other, rubbing each other, clasping trunks and trumpeting, rumbling and screaming. So much fluid streamed from their temporal glands that it ran down their chins. Moss writes: "I have no doubt even in my most scientifically rigorous moments that the elephants are experiencing joy when they find each other again. It may not be similar to human joy or even comparable, but it is elephantine joy and it plays a very important part in their whole social system." Elephantine joy can only be recognized as joy because it resembles human joy. Yet Moss is right in saying that we should not assume it is identical joy. After all, we have no idea how one feels when one's temporal glands are streaming fluid. There may be forms of joy in elephant society different from any joy that humans experience.

Biologist Lars Wilsson observed that Tuff, a beaver, looked grim when watching over her baby as it swam, and deeply unhappy if a stranger came near it, but when she was nursing it or grooming it "radiated pure maternal happiness."

A principal source of delight for many animals is their young. Certain features signal "baby animal," such as big eyes, uncertain gait, big feet, and large head. Humans respond warmly to these traits not only in baby humans but in baby animals, as well as in some adult animals. Some animals react to youthful traits with affection, others with lack of aggression or with protectiveness. Such recognition of baby features is considered to be largely innate; animals may at times feel what people feel when they say a baby is adorable. The presence of such traits in baby dinosaurs has caused paleontologist John Horner to aver that some dinosaurs must have found their babies "cute."

Tenderness may also cross the species barrier, with some animals showing distinct pleasure in caretaking. When a young sparrow crash-landed in the chimpanzee cage at the Basel Zoo, one of the apes instantly snatched it in her hand. Expecting to see the bird gobbled up, the keeper was astonished to see the chimpanzee cradle the terrified fledgling tenderly in a cupped palm, gazing at it with what seemed like delight. The other chimpanzees gathered and the bird was delicately passed from hand to hand. The last to receive the bird took it to the bars and handed it to the astounded keeper.

Another source of human happiness is pride, the feeling that we have done something well. It is unclear to what extent this can be called a self-conscious emotion and to what extent it corresponds to *funktionslust*. Lars Wilsson described the changed demeanor of Greta and Stina, when the captive beavers managed to build a dam in their enclosure. These yearling beavers had been captured as infants and had never seen a dam. Until they built theirs, they had not been particularly friendly and snarled at one another if they got too close. After the dam was built they began eating side by side, uttering friendly "talk" sounds. Not only did they no longer object to one another, they sought each other out to vocalize or groom. Greta and Stina also spent more time out of their nest box, swimming and diving in the water that their dam had deepened. The pride in accomplishment also seemed to have created friendship.

One observer described wild beavers whose dam had been severely damaged by human vandals at a season when material to repair it was hard to get. The observer arranged for suitable branches to be deposited in the pond while the beavers were asleep. The male of the pair was removing wood from his lodge to transfer to the dam when he discovered the branches. He began to swim among the branches, sniffing them and uttering loud, excited cries. One observer thought that the beaver was "rejoicing," the other that he was "marveling," but then, coming to their scientific senses, they agreed that the beaver's "subjective feelings . . . were beyond our power to ascertain."

Some captive animals experience little joy in life. For some, performing may be a chance to work, to display prowess, to feel proud. A tiger who cannot hunt, cannot mate with other tigers, and cannot explore and survey its territory has little chance of feeling pride. Perhaps, for some tigers, a chance to jump through a flaming hoop is better than nothing. But why should tigers have to settle for something that is better than nothing? To turn these magnificent animals into slaves, and then degrade them further by making them perform tricks for human amusement shows as much about human abasement as it does about animal capacities. That a tiger is condemned to slow death by boredom unless it finds pleasure in performing is a sad commentary on what humans have done to these magnificent predators.

The results of this distorted behavior affect the animals and trainers alike. Animal trainer Gunther Gebel-Williams had a tiger named India in his act for over twenty years. When he felt she was too old and deserved a rest, he stopped using her in the act, but every time he passed her cage while bringing the other tigers into the ring, she "cried." Gebel-Williams felt so sorry for India that he put her back in the act, with unfortunate results, as she was later attacked and injured by another of the tigers. Her performance may have been a source of pride, and hence greater happiness than being caged into forced retirement, but that it was the only one she had was hardly of her own choice. Along with the pride, it is important to recognize the loss of dignity. If animal dignity is little documented, it may be because the history of human dealings with animals gives little occasion for its display. The human sense of animals as "lower" by definition provides little sense of its loss.

Arguing that dolphins may be on the verge of "accepting domestication," Karen Pryor says that they enjoy performing tasks humans set for them. "I have seen a dolphin, striving to master an athletically difficult trick, actually refuse to eat its 'reward' fish until it got the stunt right." It is difficult to maintain that the dolphin "enjoys" the challenge, unless we know what its alternatives are. Would a wild dolphin ever find pleasure in such a task? Perhaps this incident speaks for dolphins having a notion of justice, or just reward, but it is difficult to say. Similar behavior in wild dolphins would be more resonant and tell us more about their society.

Horse trainers commonly observe that some horses feel pride. Secretariat, who won the Kentucky Derby in 1973, was said to be proud. As evidence it was noted by trainers and jockeys that he refused to run unless he was allowed to run the race his way—to use his burst of speed early or late in a race, as he chose—despite the fact that he was usually a docile, biddable horse. When asked whether a dog that performs well in an obedience competition is proud of itself, animal trainer Ralph Dennard replied cautiously, "It seems like it's proud of itself. They look like they're proud. They're confident; they're happy; they stand up there," and he threw out his chest as such a dog might.

Mike Del Ross of Guide Dogs for the Blind describes the gradual development of pride in dogs being trained for guide work. During the early stages, many dogs become uncertain of themselves. "It's as if they're thinking 'This is way too hard. I can't do this.'" The eyes of such dogs widen, giving them an overwhelmed look. They may lie down, go into a corner, or even huddle up in a ball. "If you don't pick *that* up right away you'll lose that dog." But if the trainer backs up, gives the dog a break from work, lets it shake off its tension, and then walks the dog through the task (which can be something as simple as leading in a straight line), the dog can regain confidence. As the dog masters what it's being

asked to do, "All of a sudden their work gets a lot less shaky. . . .
Everything comes together for them." The body language of these
dogs speaks of their self-confidence and pride. "Then in the end
they realize, 'I *can* do this!' and they enjoy it. They're proud of
themselves."

One Guide Dog trainee appeared to take pride in an accomplishment she was not taught. The dogs were housed in separate
stalls opening onto a large run. Each morning when the trainers
arrived, the dogs were let out into this area. A young German
shepherd learned to flip the horseshoe latches on the stalls. Every
morning she would let herself out and go from stall to stall letting
the other dogs out. The horseshoe latches were replaced with leash
hooks, and she learned to open those too. Finally the stalls were
fastened with leather straps, and the shepherd's efforts were
thwarted, but kennel supervisor Kathy Finger smiles at the thought
of the dog's delight when she could still open the gates. "She was
so proud of herself. She came running to us, as happy as could be,
wagging her tail . . ."

People experience territorial pride; perhaps animals can too.
The chimpanzee colony of which Washoe is a member recently
moved into a large, new facility, with indoor and outdoor exercise
areas. When one of Washoe's human companions visited for the
first time since the move, Washoe took her by the hand and led her
from room to room, carefully showing her every nook and cranny.
Washoe may have had other motivations, but she may simply have
been proud of her spacious new quarters. Sharing her place, and
her happiness in it, was also a sign of her feelings of friendship.

Reveling in Freedom

Freedom gives joy. Zookeepers, scientists who experiment
with animals, and others with entrenched interests often argue that
if all an animal's needs are met, it will not care whether it is free or
not. But many well-fed, well-treated captive animals regularly try
to escape over and over again. Freedom is relative. In spring, when
the chimpanzees at the Arnhem Zoo are allowed out of their win-

ter quarters for the first time, there is a scene of exultation as they scream and hoot, clasp and kiss one another, jump up and down and pound one another on the back. They are not free, but the additional space, the relatively greater freedom, thrills them. It looks as if it gives them joy.

George Schaller describes a two-year-old panda at a Chinese breeding center being given a rare chance to go into an outdoor enclosure. The panda burst from its darkened cage, trotted up a hill with a high-stepping gait, and somersaulted down. Again and again it raced up the hill and rolled back down. It "exploded with joy," Schaller wrote.

One of the joys of freedom is surely the ability to control one's own destiny, and a few scientists have argued that animals feel the need for such control. Zoologist J. Lee Kavanau gave white-footed mice (deer mice) the opportunity to adjust light levels in their cages by pressing a lever. He found that the mice preferred dim light to bright light or darkness, and if left alone would adjust the light level accordingly. But if he turned the lights up high the mice would frequently respond by making the cage completely dark. Conversely, if he made the cage completely dark, the mice would make the light as bright as possible. He also found that if he disturbed sleeping mice, so that they came out of their nest boxes to investigate, they would soon go back inside, but if he put them inside by hand, they would immediately come out, no matter how many times he replaced them. They cared about choice more than comfort. When given the opportunity to manage their environment, they battled fiercely for control. Because wild white-footed mice have far more control over their surroundings and activities, this matters more to a captive animal. Even if a zoo animal is supplied with all material wants, there may be something vital lacking, something it needs to be happy. One of the joys of freedom may simply be the ability to evade compulsion.

In a related vein is the story of Charles, a small octopus who was the subject of an experiment to see whether invertebrates could learn conditioned tasks as vertebrates do. With two others, Albert and Bertram, each housed in a small tank, Charles was to be trained to pull a switch so that a light went on, and then swim over to the light to be rewarded with a minute piece of fish. Albert and Bertram learned to perform this task and Charles seemed at first to be doing the same. But then Charles rebelled. He began anchoring himself to the side of the tank and yanking on the lever so fiercely that he eventually broke it. Instead of waiting under the light to receive his smidgen of fish, Charles reached out of the water, grabbed the light, and dragged it into the tank. Finally, he took to floating at the top of the tank, with his eyes above the surface, accurately squirting water at the experimenters. "The variables responsible for the maintenance and strengthening of the lamp-pulling and squirting behavior in this animal were not apparent," the experimenter noted primly.

In a captive breeding and release site for thick-billed parrots located in an Arizona forest, the captive parrots awaiting release were healthy, glossy, and well fed. Amply supplied with food, water, safety, and companionship, they resembled the best-treated pet parrots. All the same, the free-flying parrots looked startlingly better to observers. It was hard to pin down the difference. Both groups of birds had glossy plumage and bright eyes. The demeanor of the two groups probably conveyed the difference. The captive parrots were not hunched or pathetic, but the wild parrots seemed ten times better: stronger, happier, and more confident. Even as they eyed the sky for hawks, they seemed to be reveling in life. In F. Fraser Darling's classic *A Herd of Red Deer*, a similar observation is made of deer kept in paddocks compared to wild deer: something is missing.

Can animals ever be happy in captivity? Can a zoo ever be a good zoo in the sense of being a joyful zoo? Since animal behavior is so often flexible, this ought to be possible, but then most animals are not held captive by people who are asking what it would take to make the animal happy. They ask what it would take to make the animal docile, or to make a good exhibit, or to breed. The art of making zoo animals contented, thrilled, or joyous is not a subject of expertise.

Wolves breed in captivity, but it seems unlikely that a wolf who is constantly being stared at from nearby, a wolf with no place to hide, a wolf who cannot see the moon, can be a contented wolf.

It may be that a wolf cannot be happy while making a good exhibit. A raccoon may be less likely to have this problem. It may have other problems, however, no less distorting of its nature. A happy animal needs to feel safe most of the time. If it is a social animal it needs company. It needs something that it can accomplish. A dish of food three times a day may be nutritionally equal to four hours of foraging, but it is not emotionally equal.

When Indah, an orangutan, escaped from her enclosure at the San Diego Zoo in June 1993 and clambered up onto the viewing deck, she neither headed for the hills nor attacked people. She chose to go through a garbage can, put a bag on her head, taste what she found, and dump an ashtray, surrounded by an interested audience. In other words, she indulged both her curiosity about what happens on the other side of the viewing deck and her need to act on her world in her own way. This seems to indicate that in her enclosure Indah was bored. She was not fulfilled. She had things she wanted to do that she was kept from doing. In zoos, visitor after visitor remarks on the apparent boredom of so many of the animals. Many people express a sense of unease, an understanding of how they would feel under such circumstances.

An animal also needs room to wander over a range appropriate to its species. For some small animals who cherish the vicinity of a burrow or nest, an ordinary zoo cage, if appropriately complex, might be big enough. No cage is big enough for a polar bear or a cougar. Whether an animal without freedom to choose its own environment, no matter how small, can be happy is a question that needs to be asked. Is not freedom of choice basic to the meaning of happiness? It is not surprising that a favored task for captive animals to be taught is the miming of happiness. Dolphins, confined in tiny spaces, deprived of most companionship, denied the use of many of their capacities, are trained to burst into the air in a shower of spray, to dance across the surface of the water and to leap in apparent joy. The joy may even be real, but it does not reflect the overall reality of the captive animals' life.

Play

Joy often expresses itself in play, which many animals indulge in all their lives. Play, which seems to be both a sign and a source of joy, has been increasingly studied in recent years, following a long period when the subject was considered in professional circles to be less than respectable. The longtime lack of study, according to Robert Fagen of the University of Pennsylvania, was a blacklash against the work of Karl Groos, who, at the end of the nineteenth century, argued for a link between play and aesthetics, depicting play as a simplified form of artistic endeavor. Fagen notes that "the study of animal play has never quite overcome its embarrassment at Groos's attempt to unite animal psychology with aesthetics." Biologists continue to be dismayed by the lay public's interest in possible links between animal play and human creativity.

Fagen is unintimidated, and at the end of his book on play in animals, he writes:

In the play of animals we find a pure aesthetic that frankly defies science. Why kittens or puppies chase and vigorously paw at each other in reciprocal fashion without inflicting injury, repeating this behavior almost to the point of physical exhaustion, is not known. Yet this behavior fascinates, indeed enchants.

Some researchers also feel that the study of play has been neglected because play behavior has been inadequately defined. Various definitions have been propounded, more or less ponderous. Frustration is apparent in ethologist Robert Hinde's definition: "Play is a general term for activities which seem to the observer to make no immediate contribution to survival." That is, play is something done simply for the joy of doing it.

Unwieldy definitions of play (some mere lists of play behaviors) are probably better than none, however. Marc Bekoff, professor of biology at the University of Colorado, has noted a tendency among ethologists and behavioral scientists either to make definitions of difficult concepts so narrow as to be useless or to say that a concept that is difficult to define must therefore be impossible to study. "For example, some have claimed that social play is not a valid category of behavior because it is so difficult to define. While suggesting that we do away with social play by stipulatively defining it out of existence (or by defining social play as what it is not), few if any viable alternatives were offered; thus we were left with nothing!"

Play is important to animals, and, although it carries risks, since animals can be injured or killed while playing, a variety of evolutionary functions for it have been proposed. Perhaps it is a form of practice, of learning to perform tasks, theorists suggest; or perhaps it exercises developing social, neurological, or physical capacities. Cynthia Moss may have spoken for many biologists when, watching African elephants play in the rain—running, twirling, flapping ears and trunks, spraying water at each other, flailing branches, uttering loud play trumpets—she wrote in her notes, "How can one do a serious study of animals that behave this way!"

Hans Kruuk, studying spotted hyenas, complained that play "is an anthropomorphic term, negatively defined; I have merely used it as a label for some activities which in our own species would be named that way." As an example of such activities Kruuk cited four adult hyenas swimming in a river, jumping in and out, splashing and pushing one another underwater. Kruuk adds that the hyenas made a substantial detour to reach the pool.

Elephants, both Indian and African, are particularly playful. A traveling circus once pitched its tents next to a schoolyard with a set of swings. The older elephants were chained, but Norma, a young elephant, was left loose. When Norma saw children swinging she was greatly intrigued. Before long she went over, waved the children away with her trunk, backed up to a swing, and attempted to sit on it. She was notably unsuccessful, even using her tail to hold the swing in place. Finally she flung the swing about irritably and returned to her companions. The children began to swing again, and Norma had to try again. Despite trying periodically for an hour, she was never able to swing.

Norma may have been looking for entertainment because she was bored. There seems no reason to doubt that animals can be bored. Nim Chimpsky often appeared to his sign language instructors to be bored and would demand to be taken to the toilet or to go to bed when his teachers felt strongly that he was just looking for a change from his lessons, like any schoolchild.

The life of many herbivores strikes many humans as intrinsically boring. Grazing animals eat the same foods all day long every day. This would certainly bore omnivores like us, but maybe buffalo have a higher tolerance for monotony. Maybe each blade of grass seems vastly different from the blade before. Perhaps their life is a rich tapestry of excitement and intrigue, but at a sensory level too far removed from ours to be apparent. In any case, to assume that a wild buffalo is bored with its life because it would bore some humans is true anthropomorphism.

Alaskan buffalo have been seen playing on ice. One at a time, starting from a ridge above a frozen lake, the buffalo charged down to the shore and plunged onto the ice, bracing their legs so that they spun across the ice, with their tails in the air. As each buffalo skidded to a halt, it let out a loud bellow, "a kind of *gwaaa* sound" —and then awkwardly picked its way back to shore to make another run.

Animals can play in complete solitude. Bears are playful throughout their lives and will slide on snowbanks like otters— headfirst, feet first, on the stomach, on the back, while somersaulting. Two grizzly bears in the Rockies were seen to wrestle for possession of a log. The bear who triumphed lay on its back and juggled the log on its feet while roaring with delight. A quieter grizzly floated in a mountain lake on a hot day. It ducked its muzzle beneath the water to blow bubbles—and then reached out to pop them with its long claws. Tiger cubs and leopards love to jump off branches into water and will do so repeatedly. Bonobos (pygmy chimpanzees) in the San Diego Zoo play solitary games of blindman's bluff. The bonobo covers its eyes with a leaf or bag or simply puts its fingers or arm over its eyes, and then staggers about the climbing frame.

At one time the gold leaf on the domes of the Kremlin was being scratched off by hooded crows. The crows were not indulging their fabled penchant for theft. They had simply found that it was enormous fun to slide down those onion domes, and their claws did significant damage. Eventually they were driven away by a combination of recorded crow distress calls and regular patrols by tame falcons.

Animals may also play with objects. This can be seen even in some animals that are not known to play with other animals. A captive Komodo dragon in a British zoo played with a shovel, pushing it noisily about the enclosure. A meter-long wild alligator in Georgia spent forty-five minutes playing with the drops of water falling from a pipe into a pond, stalking the pipe, snapping at the drops, letting them fall on his snout and then snapping at them in midair. Captive gorillas and chimpanzees enjoy playing with dolls and spend time in other imaginative play, as when Koko the gorilla pretends to brush her teeth with a toy banana, or when the chimpanzee Loulis, playing alone, puts a board on his head, signing "That's a hat."

In other animals, object play quickly becomes social play. A captive dolphin at an oceanarium played with a feather, carrying it to an intake pipe, letting it be swept off by the current, and then chasing it. Another dolphin joined in and they are said to have taken turns. In another game three or four dolphins vied for possession of a feather, and wild dolphins play similar keep-away games with various objects. Beluga whales carry stones or seaweed on their heads and the other whales at once try to knock them off. Lions, both adults and cubs, may try to wrest pieces of bark or twigs from each other.

Teasing is a form of play, at least for the one doing the teasing. Some animals tease conspecifics; they may also tease members of other species. A captive dolphin teased a turtle by tossing it out of the water and rolling it along the bottom of the tank. Another dolphin teased a fish that lived in a rock crevice in the tank by putting bits of squid near the hole; when the fish came out to get the squid, the dolphin would snatch it away. Many people who visit captive dolphins see them mercilessly tease sea lions and seals with whom they share pools. Ravens tease peregrines by flying closer and closer past them, croaking, until the falcon lunges after them. Swans, with their large dignity, are often the target of teasing. In the water, little grebes have been seen to tweak swans' tails and then dive. On land, carrion crows may pull their tails repeatedly, leaping back each time the swan turns on them.

Foxes tease less nimble hyenas by coming close, circling, then sprinting away until the hyena can no longer ignore it and lunges for the fox. Several cases of hyenas actually catching and killing such a fox have been reported. Maybe the fox is gaining information about the hyena's powers, useful when the fox snatches bites from the hyena's kills. Maybe the fox is accustoming the hyena to its presence, also useful when pillaging kills. This gives a practical explanation for why such behavior persists, but it does not explain what the fox feels. Why should the fox not feel the mischievousness that has been imputed to the species over the centuries?

Games

Other forms of play seem to be enjoyable for all involved. Young animals and sometimes adults commonly wrestle, mock-fight, and chase one another. Sifaka lemurs lie on their backs with the soles of their hind feet pressed together and "bicycle." A favorite game of young animals of many species, from wolves to red deer, is King of the Castle, where one player occupies a high place and defends it against the onslaughts of the others. Troops of sifaka and ring-tailed lemurs often tease by barring the other troop's passage, with animals jumping in front of each other, over each other, around each other. Unlike genuine territorial disputes, the two troops mingle and bound in all directions rather than aiming for a particular goal.

To what extent animals playing games recognize implicit rules is not clear. In a few instances trainers have successfully taught formalized games to animals. A simplified version of cricket was taught to elephants of the Bertram Mills' Circus after several months of training. Elephants understand throwing objects, but batting and fielding took time to learn. It is said that after some months the elephants began to "enter into the spirit of the game" and subsequently played with great enthusiasm.

At an oceanarium, several dolphins were trained in the skills of water polo. First they learned to put a ball through a goal, each team having a different goal. Then the trainers tried to teach them to compete by keeping the other team from scoring. After three training sessions the dolphins caught on, all too well. Uninterested in strictures against foul play, the dolphins zestfully attacked one another in such an unsporting fashion that the training was discontinued and they were never again given competitive games. There is no indication that the dolphins thereafter tried to play polo on their own.

Interspecies Play

Animals sometimes find playmates across the barrier of species. In captivity, species that would be unlikely to meet in the wild are often brought together. Thus a leopard and a dog may play together, or a cat and a gorilla. One family that kept red kangaroos in their backyard with their dogs found the animals quite friendly toward each other, although there were difficulties. The dogs liked to chase and be chased by their friends, barking. The kangaroos preferred to wrestle and box, pastimes the dogs did not care for. Somehow they managed to play together.

Interspecies play, while not routine, is also found in the wild. Dwarf mongooses in Kenya have been seen attempting to play with ground squirrels, lizards, and birds. Here again, the different styles of play can form a barrier. M'Bili, a young mongoose rebuffed by her mongoose playmates, ran over to a large lizard, hopping and uttering play calls, and began tossing dead leaves about. When this produced no reaction, she danced around the lizard, tapping it and pretending to nibble at the lizard's back, forefoot, and face. The lizard closed its eyes and did not respond, and M'Bili gave up.

Another mongoose, Moja, tried to play with an African ground squirrel as he would play with another mongoose. Moja was playing with another mongoose when the squirrel hopped into their midst and, standing on its hind legs, stopped to gnaw on a nut. Moja raced over giving the "play call," rose on his hind legs, put his forepaws on the squirrel's shoulders, and started to "waltz" around with the squirrel. A playful mongoose follows this by pretending to snap at the other animal's head and neck, which Moja did, but the squirrel didn't respond, simply standing passively and letting itself be waltzed around. Then Moja pounced on the squirrel's tail and bit it, whereupon the squirrel hopped away and Moja attacked a twig instead.

Tatu, another young mongoose, had better luck with a white-headed buffalo weaverbird. Tatu chased the bird, making jumps up into the air at it. Instead of departing, the bird flew no more than a foot off the ground, repeatedly skimmed over Tatu's head, and landed on twigs close to her. Tatu was the first to tire of this game. In a still more successful match, wild river otter cubs and beaver kits have been seen playing together. Adult beavers and otters were present and paid no attention as two cubs and two kits nosed, nudged, and chased each other, both on a stream bank and in the water. The play continued until the otter parents moved their family along. Young mangabey and red-tailed monkeys, whose troops often forage together in the Tanzanian rain forest, also play-wrestle together. Interspecies play, a commonality in nature, has a special charm for humans. If two species of animals can reach out joyfully across the gap between them, it seems that humans, too, might reach across and share their joy.

Sometimes the gap between species can be very wide. Douglas Chadwick describes an old bull elephant drinking at springs in Africa. Near one pool he met a tiny blacksmith plover. Ferociously the bird lifted its wings and shrieked a threat. The huge animal departed. "While leaving, though, the old bull pranced a little bit, shaking his head, as if laughing to himself." Chadwick concedes that this description can be called anthropomorphic, that some would insist on saying that the elephant "was merely exhibiting displacement behavior, releasing a bit of tension built up in response to the bird's threat. But what is the difference between this and many occasions that cause us to shrug and shake our heads, laughing to ourselves? In fact, that is exactly what I did when blacksmith plovers stalked forward yelling at me." To refuse to see the commonality between person and elephant is to deliberately widen the gap. Another gap may come between the elephant and the bird. The elephant seems to have been amused by this encounter, but there is no reason to guess that the plover saw the joke.

Sometimes the gap may be too wide. In Bert Hölldobler and Edward O. Wilson's authoritative work on ants, we find a section titled "Ants Do Not Play" in which they refute the notion, put forward by several observers, that ants may be seen to play. The wrestling ants observed by Huber and by Stumper were not playing but in earnest, Hölldobler and Wilson argue: the contestants were members of different colonies fighting to master each other. "In short, these activities have a simple explanation having nothing to do with play. We know of no behavior in ants or any other social insects that can be construed as play or social practice behavior approaching the mammalian type." Yet nineteenth-century naturalist Henry Water Bates's description of Eciton ants in Brazil does not sound like fighting.

> I frequently saw them very leisurely employed in a way that looked like recreation. When this happened, the place was always a sunny nook in the forest. . . . Instead of pressing forward eagerly, and plundering right and left, they seemed to have been all smitten with a sudden fit of laziness. Some were walking slowly about, others were brushing their antennae with their fore feet; but the drollest sight was their cleaning one another. Here and there an ant was seen stretching forth first one leg and then another, to be brushed and washed by one or more of its comrades, who performed the task by passing the limb between the jaws and the tongue, finishing by giving the antennae a friendly wipe. . . . The actions of those ants looked like simple indulgence in idle amusement. Have these little creatures, then, an excess of energy beyond what is required for labors absolutely necessary to the welfare of their species, and do they thus expend it in mere sportiveness, like young lambs or kittens, or in idle whims like rational beings? It is probable that these hours of relaxation and cleaning are indispensable to the effective performance of their

131

harder labors; but while looking at them, the conclusion that the ants were engaged merely in play was irresistible.

Perhaps Bates is right, and some ants do play. When people think of playing with animals, we tend to think of playing with dogs or cats. It is hard to imagine playing with an ant, but this is no reason to decide that ants cannot play with one another.

There is something compelling in the recognition that other creatures enjoy play as much as we do. Jacques Cousteau wrote of whales as "sociable, affectionate, devoted, gentle, captivating, high-spirited creatures. The entire ocean is their empire—and their playground. Theirs is a 'leisure society' that predates ours by some forty million years. They spend less than a tenth of their lives looking for food and feeding. The rest of the time they spend swimming, frolicking in the waves, conversing with each other, wooing the opposite sex, and rearing their young—an inoffensive agenda if ever there was one!"

Scientists and laypeople alike have long been fascinated by the social play of canids—wolves, dogs, coyotes—because it so clearly involves a shared understanding of language and of social bonds. The play bow—when a canid lowers its forelegs to the ground and waves its tail—is a way of saying: "Everything that follows is just a game. Are you ready to play?" Dogs will attempt to play with another animal, cats, for example, but are usually disappointed in their lack of fluency in or indifference to this canid metalanguage. This gives special poignance to the play between a dog and its human friend, as if dogs recognize that they have found a companion to whom they can teach the rules. Nor do they seem unhappy at trying to figure out human rules for the games we wish to play with them. The concentrated posture a dog assumes over a stick he is waiting for his human friend to move is obviously *meant* to be slightly humorous: that is part of the game. Playing these games is almost like looking through a window into the dog's mind. We see what he intends. And the dog, too, gets a clear glimpse into our minds and knows what we want. Play, laughter, and friendship burst across the species barrier.

Rage, Dominance, and Cruelty in Peace and War

In the fifteenth century, when giraffes were known in Europe as camelopards, Cosimo de' Medici shut a giraffe in a pen with lions, bloodhounds, and fighting bulls to see which species was the most savage. As Pope Pius II looked on, the lions and dogs dozed, the bulls quietly chewed their cuds, and the giraffe huddled against the fence, shaking in fear. These leaders of men were disappointed at the absence of bloodshed, and wondered why the animals were not more savage.

History texts, today's newspapers, and our own lives testify to the fact that humans are moved by anger and hostility with regularity, despite wishes to control or at least disguise these emotions. In contrast, people are often eager to point out aggression among animals, and identify it as "animal," "brutal," or "savage." While aggression among animals is a favored topic for ethologists to study, the word *anger* is unlikely to appear in their work.

Animals do seem to get angry; they certainly do commit aggressive acts against each other, fight for turf, and hurt and kill one another. They may not do this in exactly the ways people expect, however.

Like de' Medici and Pius II, scientists have had their expectations about animal aggression confounded. Ethologists seeking to chart dominance hierarchies in groups of wild animals are sometimes frustrated when they cannot find who is dominant. They seem to suppose that with luck or labor, the true ranking among the animals will emerge. It does not always occur to them that the relations may not be hierarchical. It is as if animals at a waterhole must be arranged as neatly as academics at a granting agency.

Some people nurture the hope that animals—and if not all animals, then some cherished species—are not aggressive except in self-defense. In the society of wolves, the company of dolphins, or the murmuring of doves, only harmony is imagined. If the lion does not lie down with the lamb, perhaps at least the lambs lie down together. A look at the evidence, however, shows that while lambs may lie down together, they may very well rise to butt heads. Doves, dolphins, and wolves can treat each other very roughly. This is not to say that their whole social life is marked by conflict, only that the hope for a saintly species to become our guru of peace, love, and benevolence is unlikely to be fulfilled. Perhaps it is the expectation that is unreasonable.

Aggression spans a range from unprovoked attack to self-defense. When an animal pushes another away from food or refuses to be pushed, growls at an animal that comes near its young, or chases away a rival, it is behaving aggressively. From the viewpoint of survival, such behavior often has comparative advantages. The aggressive animal gets more to eat, keeps its offspring safer, has a better chance to mate, or faces less competition, all of which may enable it to leave more descendants. Anger and other emotions related to aggression may produce such behavior.

Of all the forms of physical force in the animal world, those most apt to be forgiven by critical humans are self-defense and defense of young. A wolf attacking a deer may be called savage and ravening, and the deer's defense may be called brave and heroic. The tigress or the bear defending her cubs is an archetype of justifiable rage. Animals like the red kangaroo, who may toss the larger of her joeys out of her pouch if she is too closely pursued, are viewed with great disfavor. Such an incident is no one's favorite animal story.

Waging War

One of the gravest charges against the human race is that only humans make war. The German writer Hans Magnus Enzensberger began his recent book on European civil war by saying, "Animals fight, but they do not wage war." We are supposed to be shamed by the fact that animals do not make war. Yet some animals do. Ant wars are the best known, but insects are sufficiently dissimilar to us that people seldom take that to heart. In recent years it has become clear that animals as closely related to us as chimpanzees can go to war. The famous chimps of Gombe attack other bands with no provocation and with deadly intent, not only patrolling their borders but making raids. They may kill and eat one another.

One account, which includes both terrorist behavior and a sudden recognition of commonality, is particularly evocative of human warfare. When a group of chimpanzees from Gombe's Kasakela group found a strange female and her infant in a tree, they barked threateningly. After a few blows had been aimed at her, there was a pause, during which some of the chimps fed in the same tree. She approached one of the males submissively and touched him, but he made no response. When she tried to leave, several of the males blocked her way. She approached another male, Satan, submissively and again touched him. His response, in an apparently xenophobic gesture, was to pick some leaves and use them to scrub the spot she had touched. Immediately several chimpanzees attacked her and grabbed her infant. For eight minutes she fought unsuccessfully for her baby, and finally escaped, badly injured. One of the Kasakela apes smashed the infant against trees and rocks and tossed her down. She was not dead, and Satan picked her up gently, groomed her, and put her down. Over the next few hours, three different males, including Satan, carried the infant tenderly, supporting and grooming her, before she was abandoned to die of her wounds. It is hard to know what to make of this strange story. Is it possible to attribute regret, a sense that they had gone too far, to these chimpanzees? Did they feel first hatred, then compassion, as warring humans sometimes do? In other encounters between bands, infants have been killed and eaten. This incident suggests mixed feelings, in which the infant went from being "enemy" to "baby."

Bands of dwarf mongooses also join battle, apparently over territory. Many are injured, and some die. One battle began with the appearance in one group's territory of a second group. Each band gathered, twittered, groomed, and marked each other with scent. Then the resident group advanced as a body and was met by the other group. The "armies" advanced and retreated, then suddenly began sinking their teeth into one another. At one point both bands retreated as if in truce, then swept back into battle. Eventually the invaders withdrew. None of the resident mongooses was killed, though toes had been bitten off, ears chewed to stumps, a tail broken. One was injured so badly that she could no longer feed herself and later died. In their next clash this same group lost.

These mass conflicts seem to be over territory. One group invades another's territory and battle is joined. Hans Kruuk observed fighting between spotted hyenas, which occurred when members of one clan killed prey in the other clan's range. Such quarrels were usually won by the residents, after prolonged threats and chasing, but sometimes the conflict escalated, and hyenas were injured or killed. Once, to Kruuk's horror, a hyena whose clan had killed a wildebeest on another clan's range was fatally injured. His attackers bit off his ears, feet, and testicles, and left him bleeding, paralyzed and partly eaten.

When Jane Goodall was able to show what looked like chimpanzees going to war, one could almost feel a sigh of relief among scientists. But compared to our own history, as biologist Richard Lewontin points out, it is the tiniest of blips, worthy of note only because it was previously unknown. We do not know, nor does Goodall pretend to know, how common it is. In many respects it feels like a man-bites-dog story, interesting because it is so uncommon. What *is* common is the overall peacefulness with which animals live together. Human history is incomparably more violent. Perhaps if we reversed our research strategy, we could find out why: A Study of Human Aggression from the Standpoint of Peacefulness Among Elephants.

Aggression over Resources

Aggression is employed by many animals to obtain access to resources like food. A principal delight of researchers on the African savannas is keeping track of which hyenas killed a wildebeest, which lions stole the carcass from them (or vice versa), and which jackals and vultures managed to snatch a bite before being driven off. Such conflict makes a dramatic spectacle. Most animals do not usually clash in this way. The wildebeest being struggled over did not, in life, stage bloody battles with other wildebeest over which of them was going to graze a patch of grass.

Competition uses a lot of energy, and many species seem to minimize such strife. In many animals there are postures of surrender that inhibit the attacker of the same species. The wolf rolls on its back, or the monkey looks away, and the attacker stops. What does an aggressor feel whose attack is checked in this way?

For many animals the creature likely to be its closest competitor, to want the same foods or the same nest sites, is another of its species, in some cases its own mate. Research suggests that size variation within some species confers survival advantage. For example, a female osprey is larger than her mate; they catch fish of different sizes, which reduces competition between them, increasing their joint supply of food.

Tame parrots often take a strong dislike to individual humans or to classes of humans, often to a whole gender. Veterinarians can grow weary of hearing clients say, "He hates all men. He must have been abused by a man in the past." Parrots have been known to conceive hatreds of all redheads, all brunettes, or all adults. While all wild-caught parrots have been abused, due to the cruelty involved in their capture and transport, this is less likely for parrots reared in captivity. But it remains unknown whether these kinds of eccentric dislikes are found in the wild.

Perhaps these parrots simply enjoy having enemies. This may promote flock solidarity, prevent interbreeding between species, strengthen the pair bond, or have some other valuable function.

Another possibility is that the irritability of parrots is related to dominance struggles in the flock. Ever since it was announced in the 1920s that chickens have "pecking orders," ethologists have been seeking and finding pecking orders—now called dominance hierarchies—everywhere. In a pecking order a chicken is dominant to some other chickens, and can peck them and push them away from food—unless it is the lowest-ranking chicken of all. And, unless it is the top bird, other chickens will in turn be dominant to it, and the chicken will allow these birds to peck it and oust it from food. The idea of dominant and submissive animals has found wide public popularity. So has the idea that aggression is valuable because it helps an animal dominate.

In recent years the idea of the dominance hierarchy has become more controversial, with some scientists asking if such hierarchies are real or a product of human expectation. It is worth noting that in wild flocks, chickens do not form rigid pecking orders as they do in poultry yards.

Some ethologists now argue that while dominance *relationships* between two animals ("the gray female is dominant to the black female") may be real, dominance *ranks* assigned to individuals ("the second-ranking female in the pack") are not. Others point out that an animal may dominate in one situation—competing to eat first—but not in another—competing for a mate. Still others say that while dominance may be important between two adult male baboons, it may not be a realistic way to describe the relationship between a female and her adolescent daughter.

By far the most serious blow to theories about dominance is the discovery that one of its most basic assumptions isn't always true. This is the assumption that male animals who are dominant are able to mate more often and produce more offspring. Such males are romanticized—by some—as potent princes of their kind, genetic heroes. But recent studies show that dominant males can't always mate more often. In the hamadryas baboon, for example, whether or not females like a male is more significant to his reproductive success than his dominance. Shirley Strum found that the more high-ranking and aggressive a male olive baboon was, the *less* likely females were to mate with him. Such males also lost out when special foods were found, apparently because they had fewer friends to share with them. Leyhausen noted long ago that when tomcats fight over a female in heat, the female is no more likely to mate with the winner than with the loser—a fact that seems to have eluded most observers. Such evidence means that many cherished theories must be reexamined. A better analysis of dominance will have to include the emotions of both dominant and subordinate animals. Increasingly, attempts to fit theories of dominance to the way animals really live seems to require such terms as *respect, authority, tolerance, deference to age,* and *leadership;* terms that begin to mix emotional concepts with those of status.

Far from trying to get mates by showing dominance, many animals try to seem anything but dominant when courting, to avoid frightening the one being wooed. A courting male mountain goat lowers his back to look smaller, keeps his horns back, and takes small steps. A male brown bear slouches, flattens his ears, takes care not to stare at the female, and acts playful.

The idea of observing animals engaged in mysterious behavior and charting a tidy hierarchy that produces testable predictions has great appeal for scientists. Sometimes the idea that hierarchies are inevitable and that this proves certain things about humans is also part of the appeal. This may be why some theorists pay more attention to aggressive species than to peaceful ones, and more attention to species where males dominate than to species where females dominate, as in the lemurs. Human interest in dominance is so great that this seems to be a particularly fertile area for errors caused by projection. Scientists' behavior may, in this respect, parallel that of recreational hunters, who often seek out the biggest males as trophy animals. These animals, likely dominant or alpha males, are not usually the tastiest animals, or the easiest to find.

Yet dominance can be a real phenomenon in animals as well as people. From human lives we recognize the drive for respect or status that may be called ambition. In a herd of scimitar-horned oryx in a wildlife preserve in the Negev Desert, a male called Napoleon had grown old and short of breath and had lost status. Rather than leave the herd, he continued to challenge other males and pursue females. His challenges were ignored, but when he pursued females the other males attacked, goring him with their yard-long horns.

The preserve managers put Napoleon in protective custody—a five-acre paddock. He escaped the next day, and was injured by another oryx. He was recaptured and treated—and escaped again. After his eighth escape from the paddock, now festooned with bolts and latches, the managers changed their approach. Since Napoleon could not be forcibly contained, they decided to give him what he wanted in the paddock. They decided that he wanted not to attack males, nor to be with females, but to be dominant. So every morning the director would enter the paddock with a bamboo pole. In ritual battle, he clattered the pole on Napoleon's horns and Napoleon threatened and charged until the director allowed the victorious oryx to drive him out. Napoleon stopped escaping and lived in the paddock until he died of old age, the top oryx of his enclosure, and apparently content. It is worth asking how an animal feels when it loses status. Do animals get depressed, do they adapt, or is it ever a relief?

Rape

Rape, a sexual form of aggression, has been observed in some animals. Biologists have observed forcible rape in orangutans, dolphins, seals, bighorn sheep, wild horses, and some birds. In Arizona, an attempted rape was observed in coatimundis (long-nosed raccoonlike animals). A large male bounded out of the bushes into a group of females and juveniles, jumped on a young female and tried to mate with her. She squealed and instantly three adult females ran snarling at the male, drove him away, and chased him for fifty yards down the canyon. In none of these species does rape appear to be the norm, but in several it does occur regularly. For instance, although white-fronted bee-eaters (tunnel-nesting African birds) form mated pairs, female bee-eaters leaving the nest must dodge males who try to force them to the ground and rape them. The males preferentially attack females who are laying eggs and thus might lay an egg fertilized by the rapist, rather than by the mate.

Among waterfowl like mallards, pintail, and teal, an unwilling female is occasionally pursued by one or more males, which can result in her death by drowning when many males try to pile on. She will fight back and flee, and her mate will try to drive off the aggressors, but their efforts at defense do not always succeed. In addition, the male of a pair of mallards will sometimes attempt to mate with the female immediately after a rape attempt by another male. These mating attempts may not be preceded by the usual mutual displays of mated pairs. In most such cases "the female visibly struggled, but in no case did she flee." The sociobiological explanation for such marital rapes is that it gives the mate's sperm a better chance of competing with the rapist's sperm. It sheds no light on how the male or female birds feel. Nor does such behavior provide any evidence whatsoever that human rape is "natural," biologically determined, or reproductively advantageous.

At one marine park, where newly captured dolphins would be given a companion who was accustomed to captivity, bottle-nosed dolphins could not be used as companions because they would torment and sometimes rape the newcomer, if it was of another species. In the wild, bottle-nosed dolphins, despite their saintly popular image, have been seen to form male gangs to sequester and rape females of their own species.

Hans Kruuk witnessed a male spotted hyena attempting to mate with a female, who drove him off each time. Her ten-month-old cub was nearby, and the male hyena repeatedly mounted it and ejaculated on it. According to Kruuk, the cub sometimes ignored this and sometimes struggled "slightly as if in play." The mother did not intervene. Yet such behavior seems to be rare in animals and accounts are difficult to find.

Anger and Aggression

In most discussions of aggression and dominance, nothing is said about anger or other emotions that might inhabit such behavior. It is very difficult to tell when anger is or is not involved in aggression. Aggressive behavior in humans can be coolly calculated: whatever motivates it does not seem to be the same thing that makes people shout and fume. There is some belief that penguins may push one of their number into the water before they all dive, to see whether leopard seals are waiting there to eat them. If this proves true, it would seem unlikely that penguins commit such aggressive acts out of anger.

Some instances of animal behavior do resonate with the human experience of anger and irritation, and perhaps are easier to understand in this way because they do not constitute stereotypical reactions. Giraffes seem to dislike cars. When a car honked at one giraffe standing in the road, the giraffe knocked the car over and kicked it vigorously. Another driver encountered two giraffes crossing a road at night, stopped, and dimmed the headlights. One giraffe got off the road but the other walked over, turned its back, and delivered a series of two-footed kicks to the radiator. To anyone who has ever been annoyed by a honking vehicle, this is the very picture of irritated behavior, even the fulfillment of a fantasy.

Karen Pryor notes that if you are teaching either people or porpoises to do a task for which you have rewarded them in the past, and then stop rewarding them, they both appear irritated: humans will grumble and look sour; the porpoises will jump out of the water and splash you from head to toe.

Pryor also describes Ola, a young false killer whale (resembling an orca), reacting to a bird called a booby. One day, during a show at the oceanarium, a booby landed next to Ola's tank. Ola stuck his head out of the water and stared at the booby. When the booby didn't move, Ola leapt up at it with gaping jaws. The booby still didn't move. By this time most of the audience was ignoring the show and watching Ola and the bird. Ola hurtled around the tank, raising big waves that splashed over the booby's feet. Still the booby didn't move. Ola submerged, took a mouthful of water, came up, and squirted it directly onto the booby. The drenched bird took flight and the audience burst out laughing. Such laughter contains an element of recognition.

Oddly, dog trainers disagree on whether dogs feel anger, although they generally agree that dogs recognize anger in humans. Mike Del Ross, an experienced dog trainer at Guide Dogs for the Blind, although confident that dogs feel fear, sadness, happiness, frustration, and other emotions, doubted that they feel anger or jealousy, even when behaving aggressively. Another Guide Dog expert, Kathy Finger, disagreed strongly, saying that dogs do indeed feel anger.

This difference of opinion may stem from differing definitions of anger, but it may also be based upon a situation that every dog trainer hears about constantly: the dog that destroys things when left alone. The owner is certain that the dog is angry at being left and is chewing furniture, digging holes, knocking things over, or barking as revenge. The trainer may be just as certain that the dog is intensely bored, and that the solution lies in better conditions for the dog, rather than in discussions of what the owner considers its unjustifiable rage. Yet the two explanations are fully compatible.

One of the colleagues of the famous Ivan Pavlov tried to discover with how much precision a dog could tell a circle from an ellipse. Food rewards were given along with the circle but not the ellipse. Each time the experimenter found (apparently by observing the flow of the dog's saliva) that the dog could tell the shapes apart, new tests would begin, with a rounder ellipse. After three weeks the dog suddenly got worse at making the distinction. "The hitherto quiet dog began to squeal in its stand, kept wriggling about, tore off with its teeth the apparatus for mechanical stimulation of the skin, and bit through the tubes connecting the animal's room with the observer, a behavior which never happened before. On being taken into the experimental room the dog now barked violently, which was also contrary to its usual custom; in short it presented all the symptoms of a condition of acute neurosis." Common sense tells us that this is not a neurotic dog, it is an angry and frustrated dog.

The difficulty of untangling anger and aggression is compounded with predatory animals, whose way of getting food is more direct than anything most people experience. (A person eating a hamburger is not thought to be gloating over the suffering of cows.) This has sometimes served as a pretext for the argument that animals are unlike humans in their savagery.

Predators are often alleged to be cruel, as is nature itself. This charge has been used to justify hunting certain species almost out of existence, like the wolf and the tiger, and is also used against smaller predators like the fox and the blue jay. Since they are cruel to one another, the argument seems to go, humans have the right or duty to exterminate them.

Cases where predators kill more than they can eat or begin eating their prey while it is still alive are viewed with particular horror. Literature arguing against the protection of wolves, for example, is full of accounts of deer "literally eaten alive" by wolves. The whistling dog (Kipling's dhole) of India, because of its short canine teeth, seldom kills its prey quickly, so is persecuted as treacherous and vicious.

It is common for some predators to begin eating their prey before it is dead. It is routine for them to kill and eat baby animals before their mothers' eyes. Does this in fact reflect cruelty? It certainly appears indifferent, lacking in empathy. In certain places and times the eating of certain live animals is considered a delicacy by humans. However, the question is not whether humans can be cruel—history demonstrates that irrefutably. The question is whether animals can be cruel.

If animals are to be excused for cruel acts like eating the baby in front of the anguished mother, on the grounds that they simply cannot understand the feelings of the other, can we then believe that they are ever kind, compassionate, or empathic, which also requires understanding the feelings of another?

Torture: The Cat and the Mouse

Cruelty covers a continuum from lack of empathy to sadism. Animals do commit cruel acts. But are they cruel? Do they torment and torture? Do they like to make others suffer? (Extremists who deny that animals can suffer must also deny that animals can be cruel, since it is not possible to delight in nonexistent suffering.) A familiar instance of an animal torturing another is the cat with the mouse. On countless occasions a well-fed cat may be seen to catch a mouse that it doesn't eat. It may not kill the mouse at once. It may, instead, toss the mouse in the air, allow it to run off and almost escape—and then pounce again. It may hold the struggling creature down with a paw, and view its desperate attempts to escape with an expression that certainly looks like one of pleasure. A leopard has been seen to play with captured jackals in the same way.

The experiments of Paul Leyhausen and others with domestic and captive wildcats show that a cat will go on chasing, catching, and killing mice long after it has ceased to be hungry. Eventually it may stop killing them but will go on chasing and catching them. Then it may stop catching them, but will still stalk them. After a great while, it may give up on mice for the time being. But the stage during which the cat is chasing and catching without killing and eating looks just like torture.

Note what the cat likes best: most of all to chase, then to catch, next to kill, least of all to eat. This may seem antithetical to survival, but this hierarchy of appetites corresponds to what a successful hunter needs to be able to do. A predator may have to chase many animals before it catches one, is not able to kill all that it catches (some prey get away), and may have to catch more than it can eat (as when it is providing for the young). It has been estimated that a tiger catches prey once in twenty tries. Many kittens practice their predatory skills with prey their mother catches but does not immediately kill. A lioness has been seen holding a live warthog in her paws while cubs looked on with fascination, and cheetah have been seen bringing live gazelles to their cubs.

Does the cat who is sated with killing mice take pleasure in their suffering? Hunters enjoy their marksmanship, their ability to find their prey. They may enjoy killing the pheasant or the deer, but most hunters would claim that they do not enjoy the suffering of the pheasant or the deer. As to cats, how can this be tested?

Consider prey that doesn't suffer. A cat can hardly take pleasure in the suffering of a ball of yarn or a wad of paper. A cat is attracted to certain attributes of prey—a scamper, an uneven gait. Mice usually show these attributes better than balls of yarn or wads of paper do. But if a wad of paper could squeak and scamper as well, it might be just as attractive to the cat. Some cats have been seen to play with paper balls while mice run around underfoot.

In addition to the movement of prey, cats are often fascinated by the idea of prey in hiding. Leyhausen reports that a captive serval (a tall, lynxlike African cat), when no longer hungry, will catch a mouse, carry it delicately over to a hole or crevice, and release it. If the mouse doesn't take the chance to hide, the serval will actually push it into the hole with its forepaw—and then try to fish it out again. This can't be good for the nerves of the mouse, but servals also play the same game with pieces of bark.

Alternatively one may ask whether cats take pleasure in the suffering of prey if it does not involve flight behavior. Would a cat enjoy seeing a mouse beaten or stretched on a rack? It seems un-likely—injured mice in traps are of only fleeting interest to cats. (If any person were to suggest actually doing such experiments, one would instantly have more data about cruelty in humans.) A cat quickly loses interest in a mouse who is too badly injured to scamper away. Perhaps the cat bats it with a paw to see if it can be induced to run again, but when it doesn't, the cat is bored. The mouse may be visibly suffering, gasping and bleeding, but if it is not trying to escape, a well-fed cat is not interested. Death itself holds no interest.

Whence comes that glee that seemed to flash over the cat's face? The cat loves to hunt, to catch, to triumph. Many predators do. They may be said to enjoy killing their prey. They are uninter-ested in how their prey feels. It is the mouse's movements, not its fear, that fascinates the cat. Catching prey is part of their work, and they enjoy being successful.

Surplus Killing

Surplus killing has enraged shepherds and poultry keepers ever since people began husbanding animals. Typically, a weasel gets into a henhouse and kills all the chickens, more than it can possibly eat; or a fox jumps a fence and kills a flock of geese, making off with only one. In the wild, orcas attack a school of fish and tear from one to another, leaving bodies floating. Bears, confronted by a river full of salmon, become more and more selective about which parts of the fish they eat, until, at times, they simply stand as if in a trance, catching and releasing salmon without even killing them. Hyenas invade a flock of gazelles at night and kill dozens, far more than the pack can eat.

Such surplus killers are charged with cruelty, evil, and wastefulness. The fact that predators may kill more than they need to eat is used to justify killing them—that is, killing animals humans do not need to eat.

Predators who surplus kill are very often animals that store food to eat later. Foxes and weasels cache food. Both wild and captive hyenas have been seen to store meat in shallow water, which keeps it from rotting as quickly as it would in air. Perhaps predators that store food have the capacity to surplus kill because sometimes the excess can be cached. And it may still surplus kill when it will be unable to store the excess, like the fox that kills a flock of geese when it can only carry off one. When hyenas surplus kill, other pack members, including pups, often eat some of the surplus. Orcas that dispatch many fish in a frenzy succeed in eating more than they would if they ate each fish when they killed it, while the rest of the school escaped. They may not estimate closely how many fish will be eaten by the pod, and so may kill extra fish.

But these are simply arguments concerning the survival value of surplus killing. On the emotional side, the question is whether some creatures *enjoy* the surplus killing. Probably some do, not because they are killing more than they need, but because they are using their abilities to the fullest, exercising their capacities; they display *funktionslust*, delight in their powers. Scientist David Macdonald, author of *Running with the Fox*, says this: "I have watched foxes surplus killing. Certainly their postures and expressions were neither aggressive nor frantic. If anything they looked playful, or perhaps merely purposeful."

Is this a case of claiming nice emotions for animals and not nasty ones? Does it seem likely that animals can be kind but never cruel? Are we, as a species, alone in our capacity for cruelty?

The Target of Cruelty

If predators are not seen to torture and delight in the suffering of their prey, they might still enjoy the suffering of one another. It would not be straying from a known pattern to hypothesize that the target of real cruelty is a creature's very nearest—its family or the members of its group. Are cats cruel to other cats? Do foxes ever take pleasure in being cruel to foxes, hyenas to hyenas? Little evidence is available. Certainly hyenas and foxes act cruelly to conspecifics at times. Even when young, littermates may attack and even kill each other. It is easy to argue that there is an ultimate evolutionary benefit to the killer, but harder to guess what the killer feels.

Perhaps such animals feel hatred, yet hatred does not seem to describe the relationship between predators and prey. From an evolutionary standpoint, rabbits would gain no advantage by hating owls, and they do not appear to do so. Fear is both more advantageous and more descriptive of their behavior. Nor do owls appear to hate rabbits. However, something like hatred does seem to exist, albeit reserved for competitors, of the same or other species. The interaction between lions and hyenas seems to be at times deeply hostile. Even when not fighting over a kill, they watch one another and attack a weakened or isolated animal. Schaller notes that a lion chasing a hyena, cheetah, or leopard does not wear the impassive face of a lion hunting. It bares its teeth and utters the calls it would use against another lion.

Animals may also hate rivals of their own species. No one is better equipped to compete with an animal than a conspecific— their needs may be identical. A wolf might not only be ousted from its position in a pack, but might be viciously attacked, driven away, or even killed by other wolves.

Congo, a chimpanzee raised by humans from an early age, became deeply unhappy when he was moved to a zoo enclosure. It was hoped that the company of female chimps would please him, but he hated them and rejected their friendly attentions. He began soliciting lighted cigarettes from zoo visitors and with these he would chase the other apes around the cage, trying to burn them. Whether he felt contempt for the nonsocialized chimpanzees or simply misdirected anger at being abandoned by his companions is unknown, yet surely his feelings were very strong. Congo's spirits continued to decline and he eventually stopped tormenting the female chimps. He died a short time afterward.

Scapegoating, the identification of one animal as a target for a group's aggression, has been seen in some animals. This is especially clear where captive animals are confined in close quarters. Leyhausen confined many cats in small enclosures to see what relationships would form. In one, the "community of twelve," two cats became pariahs for no apparent reason. If they ventured down from a retreat on a pipe near the ceiling, the other cats attacked. They dared not even come down to eat unless Leyhausen stood guard. But it is important not to succumb to the idea that such situations are inevitable, since Leyhausen found that in other "communities" no pariahs or top cats emerged. In the wild, a pariah could leave, either to be solitary or to seek another group. Yet what drives wild animals away from their group sometimes looks like cruelty. To answer the question of cruelty in animals, we will have to look harder at how they treat members of their own kind and not at how they feed themselves.

Jealousy: A "Natural" Emotion?

One source of aggression in social animals seems to be jealousy, a feeling that is often expressed as anger in humans. The evolutionary approach readily attributes a value to jealousy. Between siblings it can ensure the individual's access to food and parental care. Between mates it can ensure that both parents focus their care on their mutual offspring.

In humans jealousy is frequently repudiated. Jealous people are often admonished not to feel that way. Romantic jealousy is sometimes called unnatural, a cultural artifact. Without examining whether jealousy is wrong or unwise, it is possible to examine the statement that it is a product of human culture by asking whether animals are ever jealous. (Whether it could be a product of animal culture is a subtler question.)

Although jealousy might appear in any situation where animals gather, it is most often thought of in relation to siblings and to mates. Animal siblings can be quite vicious, going so far as to eat one another; whether this involves jealousy is not known. Group members other than siblings may be the source of jealousy. William Jordan has described what happened when the first baby was born in a group of gorillas at a zoo. The group became closer and seemed more unified—except for the mother's brother, Caesar, who was hostile toward the baby: throwing branches at her, swatting her on the head. Ultimately he climbed out of the closure "in what appears to have been a jealous snit," and was placed in another cage.

In a Swedish animal park, Bimbo, a young male elephant, was shown special attention by Tabu, an older female. When a younger calf, Mkuba, arrived, Tabu lost interest in Bimbo. Bimbo responded by surreptitiously digging Mkuba with his tusks whenever possible, and Mkuba responded to that with loud, histrionic shrieks for help from Tabu. This sort of behavior is so reminiscent of human actions that strong scientists feel compelled to take a deep breath and start numbering the animals they observe instead of naming them.

Freud formulated his notion of Oedipal jealousy in reference to humans, but Herbert Terrace interprets the chimpanzee Nim Chimpsky's actions in this light. After being taken from his mother at the age of five days, Nim was raised in a human household. His foster mother, Stephanie, observed that Nim displayed both affection and a certain hostility toward her husband. In one instance, Nim, Stephanie, and her husband were taking an afternoon nap on a large bed, with six-month-old Nim in the middle. Nim appeared to be asleep, but when Stephanie's husband put his arm around her, Nim leapt up and bit him. Terrace was unable to resist describing Nim's behavior as "downright Oedipal," though other explanations are possible.

The famous grey parrot Alex, who speaks words whose meaning he understands, is not a bird prodigy. His trainer has worked with at least one other grey parrot who learns just as quickly. Asked why Alex has learned so much more than thousands of pet parrots over the centuries, Pepperberg ascribes his success to the model/ rival method. (It may also be that some of those pet parrots did understand the meaning of words they used, but that this was not believed.) In this method, two people work with the animal, one as trainer, the other as model or rival. Thus, if Alex is to be taught the word green, the model (usually a graduate student) is shown a green object and asked what color it is. When the student says "green," the trainer praises her or him, and bestows the green object as a reward. When the student gets it wrong, there is no reward.

Alex, who has watched, is then asked to do the same task. The student may be thought of as a model, who demonstrates what is wanted and what the reward is. But the student may also represent a rival, someone who makes Alex feel jealous. Perhaps Alex doesn't really want the green object until he sees someone else get it. Perhaps he doesn't like to see someone else praised instead of him. This is currently only speculation: Irene Pepperberg's analysis of the model/rival system focuses on its referentiality, contextual applicability, and interactivity, rather than on Alex's feelings.

Parrots, who form enduring pairs, often seem jealous of their mate or desired mate. A tame parrot may suddenly become hostile to humans if it acquires a mate. Parrot behavior consultant Mattie Sue Athan estimates that a third of the calls for help she receives arise from "love triangles" when a parrot falls in love with one member of a human couple and tries to get rid of the other with displays of hostility. Orcas may also show jealousy over mating. At a California oceanarium three orcas were kept, two female and one male. When Nepo, the male, reached sexual maturity, he showed a strong preference for the female called Yaka. The other female, Kianu, repeatedly interrupted their mating by leaping out of the water and falling on them. Ultimately she attacked Yaka during a performance.

Scientists classifying animal mating have defined a number of systems in which it would be in an animal's genetic interest not to allow its partner to mate with others. They speak in terms of "monopolizing," "defending" or "guarding" mates, not in terms of love and jealousy. Yet jealous behavior, in the sense of possessiveness or enforced exclusivity in mating, can certainly have genetic effects. In the famous chimpanzee colony at the Arnhem Zoo, high-ranking males can often prevent females from mating with low-ranking males by attacking both the females and the males. Frans de Waal reports that, during the day, females may decline invitations to mate from low-ranking males. When the chimps go in at night they are put in separate cages. During this process, when the high-ranking males are caged, females have a chance to mate with low-ranking males without fear of attack, and will sometimes even rush over to their cages to mate through the bars. If it were not for the cages, the females might never dare mate with the lower-ranking males. Wild chimpanzees sometimes leave the group in pairs, in "consortships," and this may provide relief from jealous attacks.

Aggression and Nonaggression

Frans de Waal has pointed out that comparatively little study has been made of the avoidance of aggression and of peacemaking and reconciliation in animals or humans, although these are vital parts of social life. Watching chimpanzees in the Arnhem Zoo in 1975, de Waal saw one ape attack another, an altercation in which other group members immediately took part, resulting in shrieking pandemonium. There was a pause, and then the two apes who had the original quarrel embraced and kissed, while the others hooted excitedly. Pondering the episode, de Waal suddenly perceived it as reconciliation. "From that day on I noticed that emotional reunions between aggressors and victims were quite common. The phenomenon became so obvious that it was hard to imagine that it had been overlooked for so long by me and by scores of other ethologists."

De Waal has since studied reconciliation in rhesus monkeys, stump-tailed monkeys, and bonobos. Not only do these primates strive to make peace with each other after hostile encounters, but they also reconcile others who have quarreled. Mama, the oldest female in the Arnhem colony, once ended a conflict between Nikkie and Yeroen, two dominant males. She went to Nikkie and put a finger in his mouth, a reassuring gesture. Simultaneously she beckoned to Yeroen and when he came over gave him a kiss. When she moved from between them, Yeroen hugged Nikkie and their breach was over.

De Waal's argument is not that primates are unaggressive, but that the ways they handle and dispel aggression are as important as the antagonism and deserve equal attention. A full understanding of reconciliation awaits evidence on the emotions felt by the peacemakers. Similarly, we will not understand aggression, cruelty, or dominance and its attractions for animals and for humans until we understand their emotional aspects.

Compassion, Rescue, and the Altruism Debate

One evening during the rainy season in Kenya, a black rhinoceros mother and her baby came to a clearing where salt had been left out to attract animals. After licking up some of the salt the mother moved away, but the rhino calf got stuck in the deep mud. It called out, and its mother returned, sniffed it, examined it, and headed back into the forest. The calf called again, the mother returned, and so on, until the calf was exhausted. Apparently the mother rhino either could not see the problem—the calf was uninjured— or did not know what to do about it.

A group of elephants arrived at the salt lick. The mother rhino charged the elephant in the lead, who sidestepped her and went to a different salt lick a hundred feet from the baby rhino. Appeased, the mother went to forage in the woods again. An adult elephant with large tusks approached the rhino calf and ran its trunk over it. Then the elephant knelt, put its tusks under the calf, and began to lift. As it did so, the mother rhino came charging out of the woods, and so the elephant dodged away and went back to the other salt lick. Over several hours, whenever the mother rhino returned to the forest, the elephant tried to lift the young rhino out of the mud, but each time the mother rushed out protectively and the elephant retreated. Finally the elephants all moved on, leaving the rhino still mired. The next morning, as humans prepared to pry it loose, the young rhino managed to pull free from the drying mud on its own and join its waiting mother.

The elephant who tried to rescue the young rhinoceros ran some risk of being injured by the mother's attacks. Why did it bother trying to help? Clearly it would gain no genetic benefit from the survival of the rhino. Though both are pachyderms, there is no reason to imagine that elephants ever confuse rhinos with their own species. Perhaps it recognized the youth of the rhino and its predicament and felt a generous impulse to help.

Elephants can also be nasty to rhinos, even young ones. They have been seen to tease a rhino by surrounding it and kicking dust in its face. In Aberdare National Park in Kenya, a deadly encounter between elephants and rhinos took place one night in 1979. Elephants arriving at a water hole chased a male rhino away. Presently a mother rhino arrived with her baby, who began to play with one of the baby elephants. The mother elephant picked up the young rhino, threw it into the forest, and made as if to impale it with her tusks. But the mother rhino charged, and both mother and calf escaped. At this point the male rhino who had been chased away before made a return appearance. The irate mother elephant charged him, knocked him a distance of three meters, knelt on him, and stabbed him with one tusk, killing him.

There should be no more difficulty reconciling these two incidents than there is reconciling equally different human behaviors. Sometimes people behave generously toward an unfamiliar child, sometimes badly. However, while there has been no significant movement to deny that animals can fight and kill each other, there is much argument that they cannot behave altruistically toward each other, and that they lack the capacity for compassion and generosity. Yet observations of what actually happens in the real world do not confirm this view.

Young animals are often defended by unrelated animals. Other members of their group may defend them. Young white oryx will be defended not only by their mother, but by any oryx in the group. A mother Thomson's gazelle will defend her fawn from a hyena by running between them—but so will other female gazelles. Four female gazelles have been seen simultaneously "distracting" a hyena from a single fawn.

It may not be necessary for a young animal to be a member of a group to be defended by unrelated adults. Thus a researcher who was trying to mark rhino calves discovered to his dismay that the loud squeals of a young rhino bring not only its mother to its aid, but also every rhino within earshot.

When a group of chimpanzees in Gombe were hunting bushpigs and adolescent Freud captured a piglet, a sow charged and bit him to the bone. The piglet ran away, but the sow hung on to the screaming Freud. Gigi, a childless female chimp, charged the sow, who wheeled to face her. Though badly injured, Freud managed to clamber into a tree, and Gigi leapt free, escaping the pig's teeth only by inches.

Zebras energetically defend both young and grown zebras of their group from predators. Hugo van Lawick saw wild dogs chase a group of about twenty zebras, until they managed to separate a mare, foal, and yearling. As the rest of the herd vanished over a hill, the pack surrounded these three zebras. Their principal target was the foal, but the mother and yearling kept them back. After a while the wild dogs began jumping at the mare, grabbing for her upper lip, a hold that nearly immobilizes a zebra. Van Lawick thought the dogs would soon succeed and was astonished when he felt the ground shaking and looked up to see ten zebras thundering toward the scene. This herd galloped up, engulfed the three embattled zebras within its ranks, and galloped away again. The wild dogs followed for only a short distance before giving up.

Young animals are not the only ones to be defended. African buffalo sometimes defend other buffalo even when adult. A lion was struggling with an adult buffalo when several more buffalo ran up and chased away the lion and two other lions that had been waiting nearby.

Not all scientists are thrilled by such behavior. In the mid-nineteenth century, naturalist Henry Walter Bates shot a curl-crested toucan near the Amazon River for his ornithological collection. When he picked it up, he found that it was still alive, and it began to scream.

> In an instant, as if by magic, the shady nook seemed alive with these birds, although there was certainly none visible when I entered the jungle. They descended toward me, hopping from bough to bough, some of them swinging on the loops and cables of woody lianas, and all of them croaking and fluttering their wings like so many furies. If I had had a long stick in my hand, I could have knocked several of them over. After killing the wounded one, I began to prepare for obtaining more specimens and punishing the viragoes for their boldness; but the screaming of their companion having ceased, they remounted the trees, and before I could reload every one of them had disappeared.

Though Bates was in no danger from these birds, they might have been able to rescue their companion from a smaller, less well armed predator.

Consider the following claim in an article written in 1934 by the then curator of birds at the Smithsonian, writing for an audience of psychoanalysts: "There are no cases known to me of anything like compassion or mercy for the wounded in any bird . . . there are some cases on record that appear, on the surface, to be examples of sympathy or compassion for other birds. Thus, some parrots, eminently gregarious in their feeding habits, exhibit what looks like strong mutual attachment between the members of a flock. If one of their number has been killed or wounded by a hunter, the others, instead of flying away in terror, hover over the fallen one calling vociferously ('shrieking,' as some writers have put it) and themselves may fall the victims of the gunner who continues to shoot." This, the author tells us, is "not true sympathy and compassion in the human sense," but a feeling more like the "neurotic feigning behavior in birds whose nests and eggs or young are endangered." The writer's approach is psychoanalytic rather than behaviorist, yet in saying that the birds are not compassionate but neurotic, he achieves a similar denial of animal emotion.

Another form of altruism in the sense of selfless caring for others is exemplified in animals who feed or share their food with another animal, thus giving up a very tangible survival asset. As lion watchers have pointed out, old lionesses who no longer bear young and have worn or missing teeth can survive for years because the younger lions share their kills with them.

Although biologist and fox watcher David Macdonald has written that "Thou shalt not share thy food" appears to be one of the commandments of red fox behavior, he has also seen foxes bringing food to adult foxes when they are injured. One fox, Wide Eyes, was injured by a mowing machine. (Macdonald took her to a vet, who found that her injuries were fatal.) The next day her sister Big Ears brought food to the spot where Wide Eyes had been injured, uttered the whimper that summons cubs to eat (though Big Ears had no cubs), and left the food on the bloody spot where her sister had lain. Another time, a dog fox got a thorn in his paw, which became infected. The dominant vixen in his group brought him food, and he recovered.

Compassion for Illness and Injury

Tatu, a dwarf mongoose whose accidental separation from her family is described later in this chapter, injured her forepaw badly in a fight with another group of mongooses. She could no longer catch prey by pouncing with both paws. As she favored the paw and its nails grew long, making it even more unusable, she traveled slowly and lost weight. The other mongooses spent more time with Tatu, grooming her when she stopped grooming herself. They never brought her food. However, according to observer Anne Rasa, they began foraging next to Tatu at an increased rate. When they caught something, she would ask for it and they would often relinquish their food to her. As a younger, female mongoose, Tatu was "higher-ranking," so it wasn't surprising that they gave up food to her, only that they chose to forage near her so that this happened. At first, Rasa notes, she thought it was coincidence, but she soon became convinced that it was a deliberate choice on the part of the other mongooses. Although Tatu was getting almost half her food in this way, it did not prevent her eventual death. When she died, in a termite mound, the group stopped traveling, and only moved on when her body began to decompose.

In a case of compassion for less dramatic illness, a woman who was working with Koko, the signing gorilla, had indigestion one day, and asked Koko what she should do for a "sick stomach." Koko, who was given extra orange juice whenever she was ill, signed "stomach you orange." When the woman burped, Koko signed "stomach you there drink orange," with "there" referring to the refrigerator where the orange juice was kept. The woman drank some juice, told Koko she felt better, and offered her some juice. Only then did Koko indicate interest in juice for herself. Ten days later, when the same woman visited again, and gave Koko some juice, Koko offered it to her and had to be assured that the visitor felt fine before she would drink the juice herself.

Male elephants have been seen carrying young branches to an old bull elephant lying on the ground, too sick to forage for his own food.

Sick or injured animals may be helped in other ways besides feeding. As noted below, dolphins and whales will often support another member of their species and carry it to the surface if it is having trouble breathing. This is exactly what a mother dolphin does with her newborn, and what "midwife" dolphins do with a dolphin giving birth.

Animals have also been known to risk themselves for unrelated members of their species. An adult pilot whale swimming in the Pacific Ocean was shot and instantly killed by people on a ship. Its body was drifting toward the ship when two more pilot whales appeared, one on each side of the dead whale, pressed their noses on top of its head and dived with it. They managed to get it far enough away so that they were not seen again. This is particularly noteworthy because pressing down in this manner is not known to be a stereotypical cetacean behavior. Dolphins and whales have also been seen helping injured companions away from human attackers, and pushing and biting against lines fastened to nets or harpoons when others are captured.

Shooting lions with anesthetizing darts can provoke diverse behaviors, some altruistic, some not. The lions may attack neighboring lions as if they suspect them of having caused the pain. When the darted animals pass out, other lions may attack them. Sometimes they look up into a tree above them as if something has fallen on them, other times they charge the car in which the person firing is seated. They may run away for a short distance, and sometimes they climb a tree. Often they pull the dart out with their teeth, while on some occasions, other lions pull the darts out for them.

Cynthia Moss reported the case of a young female elephant with a badly crippled hind leg, broken when she was a small calf. This animal could not possibly have survived had her mother and other members of her group not made allowances for her, such as avoiding difficult terrain and always waiting for her to catch up. Gorillas, too, travel slowly to allow injured companions to keep pace. It is hard to believe that this is not a deliberate, conscious decision.

Ralph Dennard, a soft-spoken man with a military bearing, has spent nearly twenty years training hearing dogs to assist deaf people. These energetic signal dogs run to alert their owners when they hear a doorbell, telephone, timer, alarm clock, or smoke alarm. Dennard believes that dogs feel some emotions, such as fear, love, grief, and curiosity, but doubts that they feel compassion.

One family got a signal dog from Dennard to assist the father of the family. Gilly, a Border collie, joined them a few months before the birth of their second child, and they worried that she might be jealous and hostile toward the new baby. On the baby's first night home, Gilly woke the mother from a deep sleep and ran urgently back and forth between the bed and the baby's cot. The mother went to the cot and found that the one-day-old baby was silent and blue. He had choked on mucus and stopped breathing. His mother was able to clear his airway and start his breathing again. Later Gilly developed the habit of notifying the mother whenever the baby cried.

In another incident, a hearing dog woke a woman when a visiting cat jumped on the stove, accidentally filling the kitchen with gas. "Why did the dog respond to that? We don't know," said Dennard, pointing out that there was no sound—no ring or buzzer —to signal the dog to respond. Clearly, a dog might object to the smell of gas and want a human to do something about it. But in the case of the baby, what troubled the dog? The baby may have made choking sounds, but the dog had not been trained to do anything with regard to the baby. It seems clear that the dog knew the baby needed help, and wanted to summon that help. In humans, this is what it is to feel compassion.

Another signal dog, Chelsea, also showed concern for infants. Traveling with her owners on an airplane, Chelsea repeatedly tried to get them to go to the aid of a crying baby. Eventually, after a number of flights, they managed to convince Chelsea to leave crying infants to their parents' care.

An affecting account of sympathetic behavior tells of Toto, a captive chimpanzee whose owner, Cherry Kearton, fell ill with malaria. According to Kearton's account, written in 1925, Toto sat by him all day. When instructed, he would bring quinine and a glass. When Kearton asked for a book, Toto would put his finger on one book after another (there were fewer than a dozen) until Kearton indicated that Toto was touching the desired book, whereupon Toto would bring it to him. Several times during convalescence, Kearton fell asleep on his bed fully dressed, and Toto removed his boots. "It may be that some who read this book will say that friendship between an ape and a man is absurd, and that Toto, being 'only an animal,' cannot really have felt the feelings that I attribute to him," Kearton wrote. "They would not say it if they had felt his tenderness and seen his care as I felt and saw it at that time."

An ill, injured, or unhappy animal may also be offered comfort, as in the grooming of the injured Tatu described earlier in this chapter. An adult wild chimpanzee, Little Bee, was seen by researchers to climb down from a tree to bring mabungo fruit to her mother, who was too old and tired to climb the tree herself. It has already been noted that Nim, a chimp who was taught sign language, was very tender with people who wept. He was also responsive to other signs of grief. Indeed, his foster mother declared that during a time when her father was in the hospital, dying of cancer, Nim was more direct and more comforting in his response to her sorrow than was any other member of her family. Nim's thirty-sixth word was *sorry*, which he used when a companion was upset.

Compassion can occur by omission also. In one grim and inexcusable experiment, fifteen rhesus monkeys were trained to pull either of two chains to get food. After a while a new aspect was introduced: if they pulled one of the chains a monkey in an adjacent compartment would receive a powerful electric shock. Two thirds of the monkeys preferred to pull the chain that gave them food without shocking the other monkey. Two other monkeys, after seeing shock administered, refused to pull either chain. Monkeys were less likely to shock other monkeys if they knew those monkeys, and were less likely to shock other monkeys if they had been shocked themselves.

The behavior of these resisters contrasts sharply with the experiment described earlier, in which rhesus monkeys who had been reared in isolation were taped to a cruciform restraining device and put in a cage with "normal" cage-bred rhesus. The unrestrained monkeys eventually began performing various sadistic operations on the restrained monkeys. While they did touch and bite the tape, the experimenter concluded that they were not trying to free the restrained monkey, because they manipulated the tape less often than they did when there was no monkey in the restraining device. While one could argue about the restrained monkeys' lack of ability to solicit compassion, or the ability of a monkey to understand the concept of untaping another monkey, lack of compassion in one situation with one group of monkeys does not invalidate the presence of compassion under other conditions by other individuals.

The Altruism Debate

Altruism in animals has been passionately debated over the years, with a sizable school holding that its existence is impossible. Altruism in this scientific debate is not the same as altruism in everyday life. It means behavior that benefits another but reduces the altruist's ability to survive. For example, Richard Dawkins has written, "Altruism, for our purposes, may be defined as self-destructive behavior performed for the benefit of others." How could natural selection—the process by which only the fittest survive, passing on their genetic endowment successfully—ever favor an animal that wasted its energy or risked its life committing unselfish acts? It is argued that this could only benefit the animal—or rather, the animal's genes—if the one it helps is a relative. Extensive mathematical calculations have been made to show just how closely related an animal must be for it to be genetically worthwhile to help it. Altruism toward a close relative thus does not count as altruism under these rules.

Early in *The Selfish Gene*, Richard Dawkins specifies that he uses the term *altruism* to refer to behavior rather than "the psychology of motives." But behavior and motivation are not so easily separated, and to do so is to dodge an important issue. The sociobiological debate about altruism is deeply confused by the redefinition of this everyday word. If compassion for kin exists as an emotion, rather than exclusively as an adaptive behavior, then compassion for nonkin also becomes possible.

An example of chimpanzee compassion for nonkin—a person who was not suffering so very terribly—occurred when a group of the Gombe chimps were being followed by researcher Geza Teleki. He discovered that he had forgotten his lunch, and tried to knock some fruit down with a stick while the chimpanzees fed in some trees nearby. After ten minutes of unsuccessful efforts, an adolescent male chimpanzee, Sniff, collected some fruit, climbed down from the tree, and gave it to Teleki. This is altruism by any definition, since the human and the chimp were not related.

Sniff's mother died some years later, and Sniff adopted his fourteen-month-old sister, sharing food with her, taking her into his sleeping nest, and carrying her everywhere he went. Still unweaned, she could not survive without mother's milk, however, and she died after three weeks. A sociobiologist would not count Sniff's behavior as altruism, since she shared some of Sniff's genes. Yet in the usual sense of the word, a similar compassion, differing in its strength, may be said to have motivated his adoption of his sister and his gift of fruit to a hungry human.

Some acts that may appear altruistic to the ordinary person—running risks to protect one's offspring, for example—are not considered altruism in science. Reproductive success guarantees the replication of genes; if an animal does not protect its offspring, it is less likely to pass its genes along. Altruism is also discounted if aid is directed toward kin other than children. It has been shown that some animals who do not get a chance to reproduce can still ensure that their genes will be passed along by helping siblings, nieces, nephews, parents, and other relatives, since they share some genes with those relatives. Their individual fitness may not be improved, but their inclusive fitness, based on the number of their genes that survive in subsequent generations, will be increased. The more genes one has in common, the more advantageous it is, in evolutionary terms, to help a relative. This kin selection has been used to explain the existence of alloparenting, in which an animal helps to raise children not its own. A wolf who remains with his or her parents and helps raise their next litter is an alloparent. Perhaps there is no territory available for this young wolf to start its own family, so its best chance of passing its genes along lies in helping to raise its siblings, with whom it has an average of 50 percent of its genes in common. Or perhaps it is just a caring wolf, helping out the family.

Dawkins's calculations aim to predict whether altruism toward kin will take place. For example, he describes a hypothetical animal finding a clump of mushrooms and debating whether to give a food call that will attract its brother, its cousin, and an unrelated member of its own species to share the food. If it does so, it gets fewer mushrooms for itself but also benefits the brother and cousin, who share some of its genes. Dawkins's equation for the advantage of summoning the animal's relatives entails an elaborate cost-benefit calculation. It comes as no surprise to find that Dawkins does not suggest that any animal actually makes such a computation. What he does suggest is that "the gene pool becomes filled with genes which influence bodies in such a way that they behave as if they had made such calculations."

In another example, he discusses whether a male elephant seal should attack another male who has access to many female elephant seals, or whether he should wait for a more favorable time to attack. After imagining the elephant seal's internal debate on this subject, Dawkins says:

> This subjective soliloquy is just a way of pointing out that the decision whether or not to fight should ideally be preceded by a complex, if unconscious, "cost-benefit" calculation. . . . It is important to realize that we are not thinking of the strategy as being consciously worked out by the individual. Remember that we are picturing the animal as a robot survival machine with a pre-programmed computer controlling the muscles. To write the strategy out as a set of simple instructions in English is just a convenient way for us to think about it. By some unspecified mechanism the animal behaves as if he were following these instructions.

But picturing the animal as a "robot survival machine" seems perverse. Clearly it is a living, feeling creature. Arguably, this "unspecified mechanism" includes emotions.

In both people and animals altruism is likely to be accompanied by emotions, which accordingly must be examined along with it. In the case of altruistic behavior, this "mechanism" includes the altruistic emotions of compassion, empathy, and generosity. These emotions, even if they serve "selfish genes," may also bring into being genuine altruism in the usual sense.

When discussing altruism, a hypothetical situation theorists frequently use is that of saving others from drowning. In a discussion of how a gene for saving relatives from drowning might spread, biologist J. B. S. Haldane noted that he had twice saved (possibly) drowning people without ever pausing to consider whether doing so was genetically beneficial. Richard Dawkins notes: "Just as we may use a slide rule without appreciating that we are, in effect, using logarithms, so an animal may be pre-programmed in such a way that it behaves *as if* it had made a complicated calculation."

In real life, do animals save unrelated others from drowning? There are ancient tales of dolphins saving humans from drowning, but although some of these are plausible, none are documented. They seem plausible in part because dolphins and whales not only support other cetaceans but also carry inanimate objects on their heads from time to time. When they do so, they act like a dolphin mother with a baby. When a calf dies, mother cetaceans may support its body at the surface for as long as several days. Scientists who have seen female belugas supporting logs or other driftwood on their heads in this way believe that these are mothers whose calves have recently died. An Atlantic bottle-nosed dolphin carried a dead leopard shark on her snout for eight days. Perhaps it is species vanity, but it seems possible that humans are at least as winsome as logs or dead sharks.

Washoe, the famous chimpanzee who was the first to be taught sign language, lived for a time on a "chimp island" at a research institute. When she was seven or eight years old, a chimpanzee who had just arrived at the institute was placed on the island, but became panicky, jumped over an electric fence, and fell into the moat with a tremendous splash. As researcher Roger Fouts ran to the scene, intending to dive in and rescue her (a risky endeavor, considering how much stronger chimps are than humans), he saw Washoe run to the fence, leap over, land on a narrow strip of bank at its base, edge out into the mud, and clinging to the grass with one hand, pull the chimpanzee to safety with the other. Fouts notes that the two chimpanzees were not acquainted.

Asked whether he was surprised at Washoe's actions, Fouts paused, bemused. "Only later, when that theory came out and people said that there's no such thing as altruism. But prior to that . . ." He burst out, "You know, I was about to do the same thing. I didn't know the chimp that well either, and I was headed down into the water, too, taking my wallet out of my pants and getting ready to go in after her. Washoe beat me to it. So I guess I was responding to the same stimuli Washoe was—individual in trouble." Unfortunately, how the other chimpanzee behaved toward Washoe after being rescued is unknown. Fouts also cited a case in which an adult chimpanzee at the Detroit Zoo fell into a moat. The keepers were afraid to go in after him because adult chimpanzees are so strong, but a zoo visitor leaped in and saved the ape.

To help others, usually one must be able to realize that they need help. This recognition might be instinctive, cognitive, or both. One night in an arctic bay where beluga whales gather, three belugas got trapped near land by low tide. A gravel bar that they had swum over at high tide now barred their way. The three belugas, one adult and two juveniles, "screamed and groaned and trilled." The other belugas, free, swam back and forth on their side of the gravel bar and answered. A biologist waded out to the gravel bar, which would ordinarily have caused the whales to flee. This time, in their excitement, they paid no attention.

The whales were unable to help their fellows, but whale watchers kept the stranded belugas wet and they swam away on the next high tide. This is a story of raw emotion. The trapped whales were frightened and called for help. The other whales were concerned and came to help or possibly to show their concern. Whales are able to ask for aid from each other and to receive it in some situations. Here, the whales seemed to feel fear and empathy, even though the free whales could not actually help the trapped ones. Biologists Kenneth Norris and Richard Connor remark, "If . . . stories of dolphins pushing humans ashore are true, they must be viewed in the same context as humans pushing stranded dolphins back to sea."

Animals also have a capacity for dispassion, which tends to dismay people. They regularly do things that shock us, such as eat their dead infants or allow their offspring to eat one another. A lion who has lost all but one of her litter will often abandon her remaining cub. A parent who has been vigorously defending its young from a predator may, if the predator finally succeeds in catching the young one, walk away with apparent indifference— although it could be despair.

Yet just as kindness and cruelty can coexist, so can compassion and dispassion. Incidents that suggest the presence of one do not cancel out the other.

Compassion Across Species

As with any social emotion, a creature is most likely to display compassion toward a member of its own species. Some animals seem to recognize relationships broader than "member of my species," such as "fellow cat," "fellow bird," or "fellow cetacean." For many humans, the greatest thrill an animal can give is to treat us as one of their own. A remarkable instance of fellow-feeling is displayed by orcas, also called killer whales. In contrast to great white sharks, there is no known incident of an orca fatally attacking a human in the wild, though these carnivores eat anything in the sea from large fish to giant whales, to dolphins, seals, and birds, and even occasionally a polar bear. Though they could easily prey on people, they have consistently refrained from doing so. Given what they do eat, not eating humans suggests real fellow-feeling. Is this restraint compassion? Does it entail recognition of commonality? If so, our species has not reciprocated in kind.

Despite our differences, dolphins often treat humans as peers on some level. Even wild dolphins are sometimes interested in playing with humans. The famous wild dolphins of Australia's Monkey Mia Beach have been coming for years to play with people. While it is traditional to offer them fish, the dolphins often do not accept the fish, or accept but do not eat them. It seems logical enough that a creature that can easily catch fresh fish might not be tempted by an hours-dead specimen.

What goes through the mind of a dolphin when it accepts a dead fish and then lets it drift away? Two reporters who visited Monkey Mia saw a dolphin receive a fish from a tourist, then push it toward them. Confused, they accepted it. As the dolphin watched them, they felt socially awkward and wondered whether they were supposed to eat the thing, give it back, or do something else. As they dithered, the dolphin swam close, grabbed the fish back, and dived away, leaving them feeling they had committed an unknown faux pas.

Another way an animal might treat a human as one of their own would be to ask it for help. The act of asking for help—soliciting compassion—may itself be said to indicate a capacity for compassion in that species. How could an animal ask for compassion to be shown if it did not know what compassion was? Why would it have an innate ability to solicit something nonexistent in its species? Mike Tomkies tells of rescuing a wounded badger: "It explained her living alone in the sett, and perhaps why, after getting our scent and somehow sensing we were friendly, she had come close to us. It was odd how many sick wild creatures, including dying red deer in winter, came close to us, as if knowing they would be protected."

In Hope Ryden's *Lily Pond*, she recounts how an elderly female beaver she had been observing for several years, run down and with a paw injury, unexpectedly approached her. As Ryden sat on the banks of the pond with her binoculars, the old beaver Lily swam over, hauled herself out of the water, clambered up the bank, looked Ryden in the eye, and uttered the wheedling sounds of a beaver kit. Ryden's response was to begin bringing aspen branches (much appreciated by beavers) to the pond to supplement Lily's diet. The aspen was accepted. Though Ryden had brought branches to the pond before, she had done so surreptitiously, intending that the beavers not know she was the source. Lily's grown son Huckleberry was perhaps less compassionate, frequently trying to steal aspen branches from his mother's grasp.

Cynthia Moss has written about a very ill wild elephant who walked up to the window of her Land Rover and stood there, "lifting her eyelids from time to time and looking in at me. I do not know what she was doing, but I sensed that she was somehow trying to communicate her distress to me and I was very touched and disturbed." Barry Lopez, author of *Of Wolves and Men*, tells of a hunter who caught a large black wolf in a leghold trap. When he came up to the trap, the hunter reported, the wolf extended his trapped foot to the hunter and whined.

At times the appearance of a bid for compassion may be deceiving. A rabbit in extremis, as in the jaws of a coyote, utters a surprisingly loud fear scream. Other rabbits ignore the scream, neither speeding to see what is the matter nor taking cover themselves. The benefit of the scream, it is believed, lies in attracting other predators to the scene, and rabbit fear screams do attract predators. Apparently, in the ensuing fracas between predators, rabbits sometimes escape.

An aspect of empathy that is not considered genetically objectionable is that which results in cooperation, a situation in which both parties win. Thus if a lion understands that another lion is hunting a group of wildebeest, and joins in or helps, and shares in the resulting kill, this is considered cooperation, not altruism. In fact, it seems that when lions hunt cooperatively, they catch substantially more prey than they do when they hunt singly. If the lion helped another lion hunt, and then did not share in the kill, and the other lion was not its offspring or close relative, that would be considered altruism.

Compassion for One's Own

Unfortunately for the study of altruism toward kin and kin selection, observers usually have a frustrating time figuring out who is related to whom. Most scientists, when they see one wild animal aid another, have no way of knowing if the two are related or how closely. Long-term studies like those initiated by Jane Goodall at Gombe throw some light on these matters. Most ethology studies do not have access to as much history. Even at Gombe, observers may know who a chimp's mother is, but may only be able to guess who its father is. When they do know how the animals are related, they often don't know whether the animals themselves are aware that the other is, for example, their sister or uncle. A handful of studies have shown that some animals sometimes seem to favor their kin in surprising situations. Infant pigtail macaques preferred to play with other macaques who were their half siblings, rather than unrelated macaques, even though they had never seen them before. Whether this is connected with altruistic behavior, incest avoidance, or some other function is unknown.

Assumptions about which animals are kin can be mistaken. Among mountain goats, young animals may be seen following nanny goats with younger kids, as many animals follow their mothers for a year or two. Thus observers seeing a goat group comprising a nanny, a kid, a yearling, and a two-year-old might often assume this to be a biological family. However, it has been discovered that young mountain goats often follow nannies other than their biological mothers.

Altruism toward kin has scientific respectability, since it can contribute to the survival of a creature's genes. Also respectable to humans is reciprocal altruism, in which beings do favors for others in the expectation of receiving favors in return. That this does occur in animals as well as humans has been shown. Yet people as well as animals often help those who are unlikely to help them in return. Indeed, society expects minor acts of this kind to be performed on a daily basis, and when they are not, there is much indignation.

In the theoretical model of reciprocal altruism, the two animals trading favors each derive an overall advantage. An animal who does not return favors is detected by the others and ceases to receive them. Experimenters studying reciprocal altruism taped calls of vervet monkeys—the calls they use when threatening another vervet and at the same time soliciting assistance from other vervets—and later, hiding in the bushes, played the calls of different individuals and noted how vervets responded to these solicitations. They found that vervets were most apt to respond to the calls of unrelated monkeys if they had recently groomed each other or shared other affinitive behavior. In contrast, they responded to the calls of close relatives whether or not they had recently done favors for each other. It has been suggested that the necessity for social animals to monitor their indebtedness to one another has contributed to the development of intelligence.

Gratitude

While an animal might be keeping an unemotional tally of who owes whom, such behavior might also be mediated more emotionally, involving not only love, but gratitude and grudge holding. Unfortunately, gratitude is one of the most slippery emotions to pin down, so much so that cynics sometimes claim it does not exist in people. If A does something for B, and B is subsequently very nice to A, it can be argued that B is grateful. Some will argue, however, that B hopes for more favors from A, or that B has just come to enjoy A's company, or that B is acting the way society expects. If B is a dog, the same arguments could be applied. Yet most people believe that gratitude exists because they have felt grateful themselves. Why should not animals be able to feel gratitude also?

The human history of objectivity on this subject is not impressive. Perhaps because of guilty consciences, this is one of the emotions humans would most like animals to feel—toward us. Joseph Wood Krutch has told of a letter written to a British quarterly, *The Countryman*, with news of a thankful butterfly. The reader had seen a parasitic mite clinging to a butterfly's eye and had delicately removed it. The butterfly uncoiled its tongue and licked his hand. The reader thought this was a caress of thanks. As other readers pointed out, butterflies often lick human skin, presumably for the salt. It seems unlikely that a butterfly would interpret a lick as a grateful gesture: they do not lick each other as dogs do. The chance that an insect would thank a primate with this gesture seems rather small. Ornithologists are sometimes told stories of wild birds indicating their gratitude for favors humans have done them by singing. This, too, seems improbable, since there is no reason to suppose that birds know that people enjoy birdsong, but the idea is deeply appealing.

In the Negev Desert, Salim, a Bedouin stone carver, trapped a caracal, a lynxlike desert cat, who had been raiding his chicken house. He intended to kill it but relented, and after three days he let it go. It ran off, and by the next day had killed another chicken. In the ensuing months, the caracal would often come near Salim's house in the evenings, lie on the branch of an acacia and stare at Salim, who would sit on a rock and look back. Even after the cat had killed the last of the chickens, it would come and stare at Salim. Perhaps the caracal was curious. Perhaps it was hostile toward the person who had trapped and held it captive. Perhaps it was just maintaining their connection. Perhaps it felt gratitude.

Parrot trainers sometimes try to modify the attitude of a hostile parrot by arranging for the person it dislikes to rescue the parrot from a frightening situation. Mattie Sue Athan, a parrot behavior consultant, has written about a situation in which such a rescue occurred accidentally. The parrot, a very hostile African grey living in a pet store, had rebuffed the advances of several trainers. When Athan released it from its cage, the parrot bolted down the aisle to a ferret cage. The ferret grabbed the parrot's toe in a bloody bite and hung on fiercely. The parrot shrieked in pain and terror until Athan prized the ferret off. The bird at once became tame and friendly with her. The rescue method of winning a parrot's goodwill works fairly reliably and is sometimes exploited by unscrupulous trainers in cruel ways. As for whether the rescued parrot feels gratitude toward its rescuer, or merely trust and admiration—this is the same question asked of rescued humans.

Gratitude of one animal toward another, rather than toward humans, has been documented. In the Kenya bush one evening, Tatu, a young dwarf mongoose, became separated from her family after an antelope, frightened by a dust devil, hurtled through the group. At dusk mongooses retire into a termite mound, but Tatu was on a mound fifty yards from her family, afraid to cross the intervening ground. She uttered "Where are you?" calls and trotted back and forth on her mound. Her family called back repeatedly with louder and louder "Here I am" calls, but she dared not cross. By the time it was almost dark, Tatu was hoarse, and huddled on top of the mound. Her parents and another mongoose (probably her sister) finally set off toward her, keeping under cover as much as possible, while the rest of the band watched, scanning the earth and sky for predators. When the three arrived, Tatu flung herself upon them, licking and grooming them. When she had groomed all three (first her mother, then her father, then the third mongoose), they went back to the group. Was Tatu grateful, or merely glad to see her family? Her father did something unusual when she started to groom him, rubbing his cheek glands on her, which dwarf mongooses more typically do when preparing to fight one another. Conceivably this indicated anger, and Tatu wanted to appease them.

Elizabeth Marshall Thomas says predators may express gratitude toward prey, giving the example of a group of lions who had killed a kudu. One lion took the kudu's face between his paws and tenderly and carefully licked it as he would the face of another lion. As he did so, a cub joined him and also washed the kudu's face. In another instance a puma was seen to lie down and softly pat a bighorn sheep he had just killed. Such gratitude might not be appreciated by kudus or bighorns, but that would make it no less real in the cat.

Revenge

The converse of gratitude is surely revenge. Parrots are notorious for holding grudges. It is certain that an animal can take a strong dislike to a human individual and treat it unusually aggressively. To stay on good terms with a parrot, it is best not to be the one who clips its nails or trims its beak. If one emotion is possible, why should not the other be?

Ola, a young false killer whale in an oceanarium, was accustomed to a staff of human divers working in his tank. One diver took to teasing Ola surreptitiously. Oceanarium management had their first inkling of this one day when Ola placed his snout on the man's back, pushed him to the floor of the tank, and held him there. (He was wearing diving gear, so he did not drown.) Seeking to free the diver, trainers gave Ola commands, tried to startle him with loud noises, and offered fish, to no avail. After five minutes Ola released the diver. Subsequent investigation brought out the teasing.

Gratitude and vengefulness—the tit-for-tat emotions—might prove to be mediators of reciprocal altruism. It is arguable from the evidence that animals may possess the capacity for compassionate and generous feeling, hence altruistic behavior in the usual sense, such that even if this feeling evolved for genetic advantage, it produces behavior that need not always be advantageous. Some theorists have occasionally acknowledged the possibility of nonadvantageous behavior that could suggest other forces at work. Thus Richard Dawkins, in discussing the phenomenon of monkeys adopting unrelated babies, remarks: "In most cases we should probably regard adoption, however touching it may seem, as a misfiring of a built-in rule. This is because the generous female is doing her own genes no good by caring for the orphan. She is wasting time and energy which she could be investing in the lives of her own kin, particularly future children of her own. It is presumably a mistake which happens too seldom for natural selection to have 'bothered' to change the rule by making the maternal instinct more selective." Consider the reaction to this quotation if one did not know it referred to animals. A generous female animal "making a mistake" hardly proves that generosity—and altruism—do not exist among animals. Yet its possibility usually disappears, so that in the final pages of *The Selfish Gene*, Dawkins asserts:

> It is possible that yet another unique quality of man is a capacity for genuine, disinterested altruism. . . . We can even discuss ways of deliberately cultivating and nurturing pure, disinterested altruism—something that has no place in nature, something that has never existed before in the history of the world.

A recent scientific report on food sharing in vampire bats noted that "true altruism has never been documented in nonhuman animals, presumably because such a one-way system is not evolutionarily stable." Yet the results of the study are a little different. Vampire bats share food (the blood of other animals, usually horses) with other bats in their sleeping areas. This is vital to bat survival, since without food they starve to death very quickly. A small captive colony of bats was set up to see whether they shared with kin, with friends (as in reciprocal altruism), and with strangers. Bats that had hunted successfully did indeed share with relatives and with certain friends. "Only once did it occur between strangers," the report noted. Rather than showing that vampire bats never behave altruistically, this shows that they can be altruistic, even if rarely. The researcher's interpretation is that the sharing bat strangers made a mistake.

Altruistic acts, when recorded, tend to be treated as rare exceptions unworthy of note. For some humans, many of whom are scientists, there seems to be a powerful allure to the proclamation that all the world is ruled by self-interest, proving that kindness, self-sacrifice, and generosity are at best naive, at worst suicidal. Projecting this onto animals may be one of the more major hidden examples of anthropomorphism in science. That some people may be this way does not make animals this way. Yet scientific hegemony appears at stake, in proclaiming that animal compassion, something that everybody believes from experience, is dead wrong. Proving that all behavior is ultimately selfish to the bone gives some people special pleasure. Robert Frank, author of *Passions Within Reason*, has pointed out, "The flint-eyed researcher fears no greater humiliation than to have called some action altruistic, only to have a more sophisticated colleague later demonstrate that it was self-serving. This fear surely helps account for the extraordinary volume of ink behavioral scientists have spent trying to unearth selfish motives for seemingly self-sacrificing acts." There can be no question that, for want of a better term, the "politics" of what one selects to study plays a constricting role in understanding behavior.

Humans are usually excluded from these calculations, or are discussed only after a seemingly watertight case has been built for universal creature selfishness—whereupon sociobiologists suddenly announce either that human behavior is dictated by much the same rules, or humans are a unique exception to them.

Not all scientists fall into this snare; some have discussed the possibility of a generalized capacity for altruism. Richard Connor and Kenneth Norris queried whether reciprocal altruism is found among dolphins and concluded that it is, but also that the concept is insufficient to explain altruistic dolphin behavior. They postulate the existence of generalized altruistic tendencies in dolphins: "Altruistic acts are dispensed freely and not necessarily to animals that can or will reciprocate. They need not necessarily even be confined to the species of the altruistic individual." In dolphin society, Connor and Norris point out, individuals may be aware not only of the favor-granting status of other individuals with regard to themselves, but with regard to other dolphins in general. They concur with biologist Robert Trivers that such "multiparty situations" can reward generalized altruistic behavior, since individuals may become regarded by others as cheaters (or as generous). "In this case selection may favor an individual, A, dispensing altruism to another individual, B, even when A knows that B will not recompense him fully, or at all, in the future. The eventual increase in A's inclusive fitness will come from an increased tendency of those individuals who learned of A's altruism to act altruistically toward him." Once having argued that generosity is theoretically possible in animals, it can also be argued that it is a real phenomenon in some species.

In the evolutionary approach, an animal is more likely to receive compassion from a parent than a more distant relative; from a relative than from a nonrelative; from an acquaintance than from a stranger; from a conspecific than from a member of another species. One would expect even less sympathy from a species that does not guard its eggs. Even if this is true, compassion could be an overarching emotion that can and does produce behavioral altruism, even in the sociobiological sense.

Attributing altruism to animals can be wrong, as in one interpretation of the notorious dolphin slaughter at Iki Island, Japan, where fishermen killed hundreds of dolphins to stop them from competing for fish. This killing took place each year for five years, during which the fishermen had no difficulty in rounding up the dolphins for slaughter. This was puzzling to many observers who were trying to stop the killing, for it was widely believed that dolphins were extremely intelligent—perhaps more intelligent than humans. One theory to explain this held that the dolphins were being altruistic, allowing themselves to be trapped and killed in the hope that worldwide horror at the spectacle (which received international media coverage) would cause a revulsion of feeling and lead to the protection of wild animals. They were martyrs. After five years, a group of the bottle-nosed dolphins, instead of allowing themselves to be rounded up, darted under the boats surrounding them and got away. Perhaps they tired of martyrdom.

How many animals help others? How far will an animal go to help another? How much will it risk? How many people help others, against what odds? Yad Va Shem in Israel, commemorating the Holocaust, has an Avenue of the Righteous for those non-Jews who risked their lives to save Jews from extermination. As new deeds of bravery are discovered, trees in honor of the saviors are added. What would such an orchard for animals look like? Perhaps whales sing sagas of great acts of loving sacrifice by whale cows of days gone by.

Shame, Blushing, and Hidden Secrets

Darwin argued that only humans blush. In the years since, the self-aware social emotions like shame, shyness, guilt, embarrassment, and self-consciousness—all feelings revolving around how the self is perceived by others—have usually been considered exclusively human. Yet there is evidence that many animals feel them, too, and shame may prove to be a surprisingly basic emotion.

Asked whether wild chimpanzees ever appear ashamed or embarrassed, Jane Goodall laughed. "They do, actually. In the wild you don't see that very often. The best story I know of clear embarrassment was young Freud when he was about six years old. He was showing off—really, you could only describe it as showing off—in front of Uncle Figan, who was the alpha male. Figan was trying to groom Fifi, the new baby was there, and Freud was just prancing around and shaking branches and making a real big nuisance of himself. He went up into a tall plantain tree: they have a rather weak trunk like a banana. He was swaying it to and fro, to and fro, and suddenly it snapped!—and he *crashed* onto the ground. He just happened to land very close to me. I was able to see his face, and the first thing he did when he emerged from the grass was to take a quick little glance at Figan and then he crept quietly away and began feeding. That was quite clearly a big comedown for him."

Shame is one of the most vividly remembered feelings. Recalling happiness or fear or anger, people do not usually experience the emotion again at the time of recalling it. But remembering an incident of embarrassment or shame can often bring a flood of shame sweeping back. Those who blush may blush again at a memory. In human psychology and psychotherapy, shame received little attention for years but has recently begun to be considered important. It has been called "the master emotion," which societies use to enforce norms. Guilt refers to a particular event, but shame, which is more global, is said to refer to the individual's entire being. Thus a person might feel guilt about going off their diet, and shame at being fat. Guilt can also result from a private event, while shame tends to require that others know or observe or are imagined to judge.

Some argue—not especially vigorously, as they have met little opposition—that only humans have self-conscious emotions. It is said that animals are intellectually incapable of self-consciousness —although this is generally meant to show the low level of animal intelligence rather than a lack of emotion.

But while it may seem logical to conclude that such emotions cannot exist without an intellectual comprehension of how one is viewed by other creatures, this need not be the case. There is no reason to suppose that an animal couldn't feel shame without understanding why. As Darwin noted, mental confusion is a prominent symptom of shame. "I can't think clearly in the moment of embarrassment, and I don't know anybody else who can," psychiatrist Donald Nathanson has written. Emotion can exist with or without understanding the reasons for it. One animal might be ashamed or embarrassed without being entirely conscious of the reason; another might be ashamed or embarrassed and understand the reason perfectly.

Self-consciousness denotes both emotional and intellectual states. Emotionally, it can be an uncomfortable feeling of being observed (or observing oneself)—a form of embarrassment. Intellectually, it is the reflective knowledge of one's own mind, existence, and acts—a philosophical minefield.

Mirror studies with primates have been a focus of the debate over whether animals have or can have self-awareness. A chimpanzee allowed to become familiar with mirrors appears to learn that the image is its own. If such apes are anesthetized, dabbed on the face with a dot of paint, and given a mirror upon awakening, when they see the paint in their reflection they will touch their face with their fingers, examine their fingers, and then try to remove the paint. Orangutans also learn that the image in the mirror is of themselves; so far, monkeys have not done so. To some observers, this is evidence of self-consciousness. Others have sought to prove that it is nothing of the sort. John S. Kennedy follows some critics in arguing that it is more parsimonious to assume that the chimpanzee merely "forms a point-to-point association between the movements of the mirror image and his own movements." This tortuous explanation assigns mental powers to the chimpanzee that are at least as complex as saying that it knows it is seeing itself in the mirror. Its appeal is simply that it denies the possibility of self-consciousness in a nonhuman being.

The chimpanzees Sherman and Austin, part of an ape language experiment in Atlanta, are monitored by video cameras that they have learned to use in a variety of ways. After several months of exposure to their own images on the video monitors, each ape appeared to suddenly realize that the image was of himself. They then used the monitor to observe themselves make faces, eat, or swirl water around in their mouths. Both have learned to distinguish between a live image and a taped image of themselves, by testing to see if their actions are duplicated on the screen. Sherman was using a hand mirror to guide his application of Crayola makeup one day, tired of the mirror and gestured to have the video camera aimed at his face instead. He used the image to apply the makeup and to locate and remove any that had stained his teeth. Austin made valiant attempts one day to use the monitor to look down his throat while simultaneously shining a flashlight into it.

The practice of saving face is another chimpanzee behavior that implies consciousness of self. At the zoo in Arnhem, the chimpanzee Yeroen was slightly injured in a fight with another chimp, Nikkie. To the astonishment of researchers, Yeroen spent the next week limping dramatically—but only in Nikkie's view. A chimpanzee wishing to make peace with another sometimes will not approach directly, but will pretend to discover a nonexistent object and use the resulting gathering of many chimps to make contact with the former adversary, a strategy Frans de Waal thinks is used to save face.

In a diverse range of species, evidence that animals know when they are being observed suggests self-awareness. When a male baboon yawns, his impressive canine teeth are ostentatiously displayed. Observing wild baboons, biologist Craig Packer found that males with worn or broken teeth yawn less than males with teeth in good condition—unless there are no other males around, in which case they yawn just as often. Chimpanzees have been known to avoid glancing at a food source they know about but other chimps do not. On several occasions, lions in the Serengeti who had caught prey in high grass broke with the usual lion habit of beginning to eat at once. Instead the lion sat down and glanced around for as much as five minutes, as if it had caught nothing. When other nearby lions departed, the lion began to eat. A mountain goat who sees a predator will often walk away calmly and slowly and then, the moment it is out of the predator's line of sight, take off, running at full speed. These animals act as if they were conscious of others perceiving their behavior and want to affect that perception. This level of self-consciousness might not permit a goat to look into a mirror and think, "That's me," but might show self-consciousness nonetheless. Self-awareness need not be all or nothing.

Shyness, Modesty, and Embarrassment

Embarrassment and shyness, also considered self-aware emotions, seem to be about being seen—in a bad light, or when one does not wish to be seen. Koko, the signing gorilla, has shown a rather touching form of embarrassment. Among her toys are a number of puppets and dolls. She was once seen signing "kiss" to her alligator puppet. On another occasion Koko signed "kiss" to her blue gorilla doll and "bad bad" to her pink gorilla doll. Then she signed "chase tickle," slammed the dolls together, made them wrestle, and signed "good gorilla good good." On each occasion, and other similar occasions, the moment she saw that she was being watched, she stopped playing.

Animals do not wear clothing to conceal parts of their bodies that humans in many cultures consider vital for adults to conceal. They do not hide many actions that correspond to actions people often prefer to hide. This does not necessarily mean that there is nothing they choose to hide or keep private. The courtship of one captive bird may be an example.

Alex, the verbally accomplished African grey parrot, may be imprinted on human beings. According to Irene Pepperberg, he attempts to court certain of her male students. When courting, Alex may regurgitate food and do a little ritualized dance. "If he is courting one of my students and I walk in, he immediately stops," says Pepperberg. Perhaps Alex is embarrassed. If, on the other hand, he merely wants a little privacy, why does he want it? Perhaps he is trying to avoid competition. Is he shy? Shyness is an emotion that seems to shade into fear, the fear of being seen, and it may be that shame is also related to fear.

The essence of shame is the unpleasant feeling that one appears badly—weak, stupid, dirty, helpless, or inadequate—and the dread of appearing this way. At first sight, shame need have no connection with fear. At one oceanarium (where the animals were never punished) a bottle-nosed porpoise, Wela, was trained to jump out of the water and take a fish from a person's hand. One day when this stunt was being photographed, trainer Karen Pryor was distracted and forgot to drop the fish as she usually did. As a result, when Wela grabbed the fish, she inadvertently bit Pryor's hand. Wela, appearing "hideously embarrassed," went to the bottom of the tank, put her snout in a corner, and would not come out until Pryor got in with her, petted her, and coaxed her into calmness.

Wela's behavior is comparable to that of a dog who barks and threatens someone coming into the house—and suddenly realizes that the someone is its owner. From a barking, bristling, menacing creature, the dog suddenly deflates to a wriggling, whining, tail-wagging pup. It has been argued that the comic reversal of behavior in such a dog does not mean he or she is embarrassed, only seeking to appease a dominant animal—its owner—by showing submission. Whether this is accurate or not, it does not seem to describe Wela's behavior, which looks more like embarrassment, a form of shame. The chimpanzee Washoe was seen to make a similar mistake, threatening an old friend (who had grown four or five inches since they had last met) before she recognized who it was and reacted with what would be called embarrassment if Washoe were human. One might say that such behavior is merely ritual submission, but then the same description could apply to an embarrassed human apology.

Dog trainers at Guide Dogs for the Blind say that old dogs who have lost bladder or sphincter control seem to be embarrassed or ashamed. One otherwise healthy sixteen-year-old dog in this condition refused to go indoors as she had always done.

One text on animal behavior evades acknowledging animal emotion by insisting, "It is certainly acceptable to say our misbehaving dog is acting as if embarrassed. It would be entirely without foundation to say it was embarrassed, even though we admit to the probable existence of emotions in animals."

Blushing

The blush is a primary evidence of shame in people. Charles Darwin, who investigated blushing at great length, seems to have been surrounded by people who blushed at the least provocation. He noted that it was usually accompanied by other signs of shame such as averting the eyes, the face, or the whole body. He had difficulty explaining the value of this phenomenon, and gave a rather Lamarckian explanation for it. Humans, he said, care about their personal appearance and the opinion of others. When people feel attention, especially critical attention, being directed at them, this "excite[s] into activity that part of the sensorium which receives the sensory nerves of the face; and this will react through the vaso-motor system on the facial capillaries. By frequent reiteration through numberless generations, the process will have become so habitual . . . that even a suspicion of . . . depreciation suffices to relax the capillaries, without any conscious thought about our faces."

After questioning British missionaries stationed around the world on this subject, Darwin concluded that people of all races blush, and that it is not learned, since people who have been born blind also blush. (His data contradicted those defenders of slavery who alleged that Negroes did not blush because they were incapable of shame and hence not fully human.) He called blushing "the most peculiar and the most human of all expressions. Monkeys redden from passion, but it would require an overwhelming amount of evidence to make us believe that any animal could blush."

It would have interested Darwin to know that animals other than monkeys also exhibit reddening skin. The ears of a Tasmanian devil (a small, carnivorous marsupial) in the Frankfurt Zoo turned red "during a state of excitement." Some birds blush, as can be seen on featherless areas of skin. Like the turkey, the spangled honeyeater and the smoky honeyeater have unfeathered wattles that blush "when the bird is excited." Macaws that have bare skin on their cheeks can be seen to blush. They do so when excited or enraged, and, according to parrot behaviorist Mattie Sue Athan, they have also been seen to do so if they fall accidentally while clambering down from a perch. This certainly looks like embarrassment. On the other hand the macaw might just be angry that it has fallen. Perhaps it will turn out to be true that humans are the only animals to blush self-consciously. Humans are unusually devoid of fur, feathers, and other coverings, after all, and so provide a large canvas for this effect.

On the other hand, it may be that the function of blushing is not, or not wholly, a visual one. The phenomenon of blushing need not be visible. Many people feel tingling skin—and shame—without visibly reddening. If people flushed, paled, and turned green with the frequency found in fiction, society would be a much more colorful place. Perhaps many species of animals blush unnoticed. No one has checked to see if, under the fur, a raccoon tingles with mortification or flushes with pride. Whether macaws also blush on the parts of their bodies covered by feathers or whether other parrots flush beneath their feathers is unknown. But even if they do not, it does not necessarily follow that if animals do not blush, they do not feel shame.

The Advantages of Shame

If shame proves to be widespread in the animal kingdom, the evolutionary approach would predict that it should confer some advantage. Just what might be adaptive about global self-accusation is not immediately apparent.

The self-conscious emotions seem to appear early in the lives of humans. In one series of experiments, researchers gave small children toys cunningly designed to fall apart and then videotaped their play. When a toy broke, some children cried; some looked for another toy; some appeared ashamed or guilty. Some children looked away, their bodies "collapsed" in what is considered a typical shame response. A child who appeared tense and averted its gaze, but then tried to fix the toy, was thought to be showing a guilt response. Helen Block Lewis, an early theorist in the field of shame and guilt, viewed human shame and guilt as regulators of social interactions that combat narcissism and punish transgressions of group mores. Blushing signals to other group members that the blusher recognizes such transgression and, therefore, recognizes the group rules.

Psychiatrist Donald Nathanson does not consider shame a social emotion. He cites an experiment in which three- to four-month-old infants could control a display of flashing colored lights by turning their heads. The babies apparently loved doing this and squealed with pleasure when the lights went on. When the experimenters changed the apparatus so that the babies' efforts were unsuccessful, the infants' heads and necks slumped, their breathing quickened, blood flow to their skin increased, and they turned their faces away. Nathanson and other theorists interpret this as a primitive shame response that was independent of whether or not other people were present, and hence argue that shame is not necessarily a social emotion. (It is not clear how disappointment and frustration can be excluded as possible explanations.) In Nathanson's analysis, shame is "a biological system by which the organism controls its affective output so that it will not remain interested or content when it may not be safe to do so, or so that it will not remain in affective resonance with an organism that fails to match patterns stored in memory." He believes that it evolved comparatively recently.

As for the advantages of global self-accusation, Nathanson argues: "If you were going to design a system capable of learning from experience and educating itself, you might as well build in the capacity to magnify failure. Shame augments our memory of failure and protects us from whatever danger might occur, when, in a moment of need, we might try something well beyond our capacity."

Another possibility is that shame might keep animals from attracting the attention of predators. Humans feel ashamed not only of their actual or perceived faults, but often of their differences from others, even when those differences are neutral or even positive. To be stared at can be unnerving, even when the stare is an admiring one. Often people are uncomfortable when praised. To be singled out in any way can be acutely embarrassing—and may also feel perilous.

Predators single out prey. Some predators select prey on the basis of physical condition, thus culling out sick and injured animals, as well as young animals. They may study herds of prey animals, chase some of them, and make an all-out effort to catch only a few. An examination of the bone marrow of wildebeest killed by lions revealed that a large percentage were in poor condition. Hyenas hunting make random passes at herds, or zigzag through, then stop and watch them run, switching their attention from animal to animal, apparently looking for potential weakness. One experimenter who was shooting wildebeest with anesthetic darts in order to measure and tag them found that, if he was not careful, these same animals would at once be killed by hyenas when he released them. Although they looked normal to humans, and seemed to be able to run as fast as ever, the hyenas noted some difference. He had to herd the hyenas away with his vehicle until the wildebeest had more time to recover.

Predators also notice other differences. A researcher once marked some wildebeest by painting their horns white. Within a few months, almost all of these animals had been killed by hyenas. Hans Kruuk has noted instances in which hyenas pursued animals who were presumably in good physical condition but were acting oddly and were then singled out by hyenas. At night, when dazzled by the headlights of a car, wildebeest ran in an odd way—and were instantly pursued by hyenas. Away from the headlights, the wildebeest quickly got their bearings and escaped.

Kruuk also saw a herd of several hundred wildebeest in which only one was rolling and showing territorial behavior. These actions, unremarkable in another context, instantly attracted a hyena's pursuit. The wildebeest escaped easily. That the wildebeest escaped in these instances supports the idea that the hyenas were detecting difference rather than weakness.

Schooling or flocking behavior can baffle some predators in a very simple way: by preventing them from focusing on individual prey. When a few members of a school of small silver fish were dyed blue, not only were they more frequently attacked by predators, but so were the normal silver fish next to them. Faced by a cloud of identical fish, the predator could not pick out an individual, but it was able to pick out the blue one or the one next to the blue one.

Prey animals often seem aware of the assessment of predators. In Zaire, Paul Leyhausen saw two uneasy-looking male kobs (large antelope) near a river. Presently he saw two lions lurking near the kobs, moving from behind one bush to another. The kob closest to the lions appeared to grow calmer and began grazing, but the other began to run back and forth in alarm. It soon became clear from their movements, Leyhausen says, that the lions were stalking the farther kob—and that both kobs knew this well before the human observer figured out that the lions were not going for the closest prey.

Prey animals also seem able to tell when predators are hunting and when they are bent on other business entirely, and adjust their flight distance accordingly. A wish to conceal weakness and difference—behavior resulting from fear or dislike of being scrutinized—could lead animals to take actions to avoid predation. They could pretend not to be weak, minimize their difference, or hide from the view of predators.

Predators are not the only creatures who might take advantage of an animal's display of vulnerability or weakness. Animals of the same species are all too likely to be alert to signs of such weakness and to exploit them. When lions in the Serengeti were shot with anesthetic darts, some of the other lions took the opportunity to attack them (and were driven off by the researchers). Thus, shame might motivate animals to hide weakness from members of their herd or pack. If a caribou appears visibly weak or lame, it will be the first in the herd to be attacked by wolves, but the wolf who appears weak or sick may lose status in the pack. This can have serious consequences in terms of having offspring.

To survive, then, an animal must not only be fit, it must look fit. The sense of shame, painful to experience, may provide an emotional reason to hide infirmity.

Sickness and injury are often concealed. To the despair of animal breeders and veterinarians, many captive animals will diligently conceal all signs of sickness until they are too far gone to be saved. Birds are particularly adept at this, sometimes hiding all symptoms and enduring secretly until the moment when they literally topple from their perches.

Scottish red deer leave the herd when they become sick or are injured. At one time it was suggested that they did this for the good of the herd, but it seems probable that a lone deer is less likely to be spotted by a predator than a herd of deer—and if a herd of deer is spotted, then the sick or injured one is most likely to be the first animal attacked. If the deer recovers, it returns to the herd. If predators zero in on difference, not merely weakness, animals might feel vulnerable or ashamed of things that attract the gaze of others.

It might seem odd that shame could lead to blushing. At first glance it seems counterproductive for an animal to blush visibly or physically show embarrassment: it does not look good, and isn't the objective to look good or at least not to stand out? The tingle of the blush could conceivably signal the blusher to hide (or give them something to hide), which serves the purpose of concealing the original weakness that the blusher was embarrassed about. Most animals do not blush visibly, if at all, so it may be that humans (some of us) do have a small claim to distinction in being the most visible blushers, even if we do not prove unique in feeling shame.

Pet owners often say that their cat or dog hates to be laughed at. Elephant keepers have reported that elephants who are laughed at have responded by filling their trunks with water and spraying those who are mocking them. It seems curious that animals that do not laugh might recognize and resent laughter. But perhaps laughter should be considered the equivalent expression of something they do feel themselves, and they are better translators than we are.

Guilt

Guilt—feeling remorse for a particular act—can be trickier to pin down than shame. An action that carries overtones of guilt usually does so because our culture has informed us that it is wrong. Guilt is easily confounded with fear of discovery and subsequent disapproval or punishment. The chimpanzee Nim Chimpsky, as we saw, was taught the sign for "sorry" and used it when he had "misbehaved," just as Alex the parrot has said "I'm sorry" after biting his trainer. The examples of Nim's misdeeds that are cited—breaking a toy or jumping around too much—do not seem to be things that a chimpanzee would naturally regard as bad. Like human children, he knew it was misbehavior only because he had been taught so. Sometimes Nim signed "sorry" before his teachers noticed what he had done. Whether Nim felt something other than the desire to avert possible wrath is unclear. On the other hand, this is often unclear in humans, as well.

Washoe's adopted son Loulis was teasing Roger Fouts one day, "just being a pill," and poked him harder than usual, cutting Fouts with his nail. "I made a big deal of it, crying and so on. Later, whenever I showed him that, to make him feel guilty, if you will, to use it to exploit that, he would squeeze his eyes tight and turn away. He would refuse to look at me whenever I tried to show him or talk about this old, old scratch that he had given me." There are a variety of possible interpretations for this extraordinarily familiar behavior, but guilt is very strongly suggested.

Dogs are the most familiar guilty animal for most people. Desmond Morris has argued persuasively that dogs do feel remorse for their actions at times. When a dog that has committed some misdeed greets a human in an unusually submissive manner before the person has any reason to guess what has happened, Morris says, it cannot be getting cues from human behavior. "It has an understanding that it has done something 'wrong.'"

Human shame has only recently been deemed a respectable field of study. Donald Nathanson recounts how, early in his career, he organized a symposium on shame. When it was over a friend took him aside, complimented him on the success of the symposium, and urged him not to do any more work on shame, lest he get a reputation for it. "It was in that moment that I learned that the very idea of shame is embarrassing to most people," Nathanson says. These are emotions to be hidden. Perhaps animals have been successful in hiding them from our gaze.

If social animals feel guilt and shame, other animals might learn to take advantage of this. When a young chimpanzee makes a fool of itself, as Freud did in the incident recounted by Jane Goodall, it would seem possible for other chimpanzees to ridicule the young animal, to draw the attention of their companions to the situation, and to exaggerate their reactions. There does not seem to be any evidence that they do in fact mock each other this way, and if so, in this they differ from human animals.

Beauty, the Bears, and the Setting Sun

One afternoon a student observing chimpanzees at the Gombe Reserve took a break and climbed to the top of a ridge to watch the sun set over Lake Tanganyika. As the student, Geza Teleki, watched, he noticed first one and then a second chimpanzee climbing up toward him. The two adult males were not together and saw each other only when they reached the top of the ridge. They did not see Teleki. The apes greeted each other with pants, clasping hands, and sat down together. In silence Teleki and the chimpanzees watched the sun set and twilight fall.

The sense of beauty is not usually defined as an emotion. Yet it does not seem to be a wholly intellectual experience. Sometimes beauty makes people happy, sometimes sad; perhaps the experience is partly cognitive and partly emotional. Human beings have certainly always preferred to reserve the appreciation of it exclusively for our own species.

The chimpanzees who watched the sundown with Geza Teleki were not unique. The primatologist Adriaan Kortlandt recorded a wild chimpanzee gazing for a full fifteen minutes at a particularly spectacular sunset until darkness fell. Some who have observed bears in the wild speak of them sitting on their haunches at sunset, gazing at it, seemingly lost in meditation. From all appearances it would seem that the bears are enjoying the sunset, taking pleasure in the aesthetic experience. Scientists laugh at the naïveté of this interpretation. How could a bear be capable of aesthetic appreciation, a contemplative state? Some aesthetes also believe other people are incapable of such a state, or such a refined one; many nineteenth-century scientists claimed that "lower" races could not enjoy the same aesthetic emotions that they (as members of "higher" races) could. One can doubtless go too far with this, listening, for example, to a bear exhaling and claiming that he is sighing with melancholy awareness of the transience of things, observing his world and thinking that one day he will no longer be present to witness such beauty—an ursine Rilke. Momentary awareness of his own mortality is almost impossible to prove; a sense of beauty is easier.

Why should any creature have a sense of beauty? Some have claimed that human artistic creativity is rooted in exploratory play. Perhaps a sense of beauty rewards us for making our way through the world and for turning our senses upon it. Finding one's children and other loved ones beautiful is valuable. In a broader sense we may have evolved to see the world around us as beautiful, to enjoy gazing on it, listening to it, breathing it in, moving through it, feeling and tasting it. Perhaps it serves no useful purpose beyond the satisfaction it inherently provides, a value in itself.

Before an animal can find beauty in some set of sensations, sounds, or images, it must be possible for the animal to detect these things physically, to sense them and to perceive them. Animal senses are poorly understood but seem to be tremendously variable.

It is often said that animals, or some of them, are color-blind. This strangely counterintuitive assertion has been widely repeated as scientific fact for decades and has made its way into some textbooks. Despite numerous articles in the scientific and popular press reporting color vision in animals (including dogs), the idea that animals can see in color is still frequently described as a "myth."

Humans do have excellent color vision. Along with many

other primates, we are classified by optical scientists as trichromats, which means that we construct the range of colors we see from three basic colors. A person and an ape may see precisely the same colors when looking at a setting sun. Many mammals, including cats and dogs, are dichromats, using two basic colors. They see in color, though not as diversely as people. A few nocturnal animals, like rats, may be color-blind. Some birds use four or five basic colors; perhaps their color vision is better than ours. It has been known for decades that some insects can see ultraviolet light; recently it has been discovered that some birds, fish, and mammals can too. At least one species of bird, the Australian silvereye, can apparently "see"—that is, detect with the eye—magnetic fields.

If animals were without color vision, of what benefit would be a baboon's brightly colored face and rump, or a peacock's tail? Yet even among those who concede the peafowl's capacity to detect plumage colors, there are some who argue that it is unlikely to *appreciate* those colors. A recent natural history book for popular audiences has it:

> What is it about the peacock's fan that puts the female in the mood for mating: The iridescence? The graceful shape? The spots that look like eyes? The truth is, what wows humans may not impress female peafowl at all. Instead, the sheer size of the fan may be the irresistible feature because of what it says about the bird that carried it. Namely, if a bird with a large fan has survived to breeding age despite his unwieldy handicap, he must be both strong and wily. In the same way, females may value colorful plumage not for its beauty, but because the sheen shows that the bird is free from parasites. Females that can spot these superior traits are rewarded by passing on their genes to offspring that, like their "handicapped" father, are more likely to survive and reproduce.

How seriously should this be taken? The view that the peahen *cannot*, on the one hand, be drawn to beauty, but *can*, on the other, think: "Shiny feathers mean low parasite load—I'll mate with this one so my chicks can benefit from his genes," is untenable. If any such claim for a peahen's ability to draw intellectual conclusions were explicitly made, it would promptly be rejected. Yet the assertion that the peahen does not sense beauty is in line with assertions routinely made in the field of animal behavior.

If the peahen is not thought of as a calculating gene-shopper, what can an evolutionary approach suggest is going on? If the proximate cause is that she admires the peacock's tail because she finds it beautiful—and in humans it takes neither a powerful intellect nor extensive aesthetic training to do this—then she may mate with him, which has the ultimate result of selecting the male with the best genes. While human beings occasionally refer to others in terms of their genetic potential, this is not what is typically thought to go through the mind of a person smitten with either lust or love.

To return to the question of animal senses, acute hearing is usually conceded to some animals. Nonetheless it took Karl von Frisch, more renowned for his discoveries about bee language, to show that fish hear, in an article with the title "A Dwarf Catfish Who Comes When You Whistle to Him." Lately it has been discovered that the hearing of an unknown number of species goes far beyond our own: elephants communicate extensively in sounds too low for us to hear, and shrews, like bats, practice echolocation by means of sounds too high for human ears. In the case of birds, the implications of this are not always realized. Birds are about ten times better than humans at discriminating sounds temporally. Thus, in an interval in which we hear one note, a bird can hear ten notes. When recordings of birdsongs are played at reduced speed, it is found that they use this ability—in songs that often contain sequences of notes that pass by too quickly for the human ear. A blackbird's song that sounds like a rusty hinge to us may sound quite different to the blackbird. When people listen to or imitate birdsong, they may be overlooking great complexity.

Pet birds often appear to enjoy human music. They may prefer certain kinds of music or react differently depending on what music is played. Gerald Durrell has written of a pet pigeon who listened quietly to most music and snuggled against the gramophone. When marches were played, he would stamp back and forth, cooing loudly; to waltzes, he would twist and bow, cooing softly. Grey parrots sometimes flap their wings in delight when they hear favorite songs. Given the difference in auditory acuity, one wonders whether much human music might not sound slow and sepulchral to birds or whether they hear sounds from the instruments that the musicians are unaware of making.

Numerous species of animals utter lengthy and complicated calls that humans enjoy hearing. It would be odd if humpback whales did not appreciate their own songs or if wolves did not like the sound of their howling. For this to be true one must imagine that all the care with which a humpback composes, performs, and alters its song has no positive or negative import, only a communicative function toward which the whale feels nothing and that whales listen to other whale songs only to extract data from them. This view paints whales as far more cerebral creatures than people, all mind and no heart. Canid howling is not a random procedure; as anyone who has howled with a dog knows, dogs adjust their howling according to other sounds they hear. Hope Ryden observed a pair of howling coyotes who never howled on the same note. When the male's howl hit a note the female was on, she at once dropped her pitch; when she howled on his note, the male instantly switched to falsetto. Such duets are thought to convey the information to other coyotes that there are two coyotes howling, not just one, thus indicating that a territorial pair is present. This seems likely, but does not mean that the coyotes do not feel that the howling sounds better that way. There is no reason why the mechanism by which this advantageous behavior is obtained could not be aesthetic appreciation of song.

Similarly, the calls of gibbons seem to serve a territorial function, yet also might rise from or display an aesthetic sensibility. Gibbons sing together daily. In most but not all species, male and female calls are different. The duets of pairs, in which they exchange notes, may be sung spontaneously or in response to the songs of other gibbons. In most gibbon species, males sing long solos and females utter ringing "great calls." Juveniles often join in.

Jim Nollman, whose avocation is playing music to and with wild animals, went to Panama in 1983 to try to make music with howler monkeys, who live in family groups and call extensively. Nollman writes that a zoologist (whose study of the howlers spanned a decade) predicted that they would not be interested in his playing, except perhaps to utter a few howls as a territorial claim. Finding a tree with howlers in it, Nollman sat beneath it and played his shakuhachi flute. The entire family at first responded with loud howls. Then one monkey began to howl between the notes of the flute, in apparent response. After an hour, darkness ended the interchange. In following days, the family did not howl with the flutist's playing, but instead descended to low branches and watched him intently, despite their reputation for shyness. Whatever the howlers thought of Nollman's flute music, it seems clear that they found it absorbing although they were aware it was not produced by one of their own. Perhaps they liked it. Even if they disliked it, that might represent an aesthetic opinion.

Michael, a gorilla in a sign-language program, is fond of music, and enjoys the singing of tenor Luciano Pavarotti so much that he has been known to refuse an opportunity to go outdoors when a Pavarotti performance is on television. He likes to tap on pipes and strum on strings from burlap bags. Unfortunately, Michael is so strong that he would have a hard time not destroying a musical instrument.

What role, if any, the pleasures of taste play in animal eating is almost wholly unknown. Siri, an Indian elephant confined in a small zoo exhibit, was often seen to step delicately on an apple or orange, split it open, and then rub the pieces into her hay. Her keeper believed that Siri did this to flavor the hay. A wild elephant eats a wide variety of plants, which presumably have differing tastes. The diet of captivity is far more monotonous. Most animals, like us, seem to dislike bitter flavors and enjoy sweet ones, a discrimination that has survival value in some instances. In such simple distinctions may lie the beginnings of aesthetics.

Compared with many animals, humans apprehend little of the sensory realm of smell. We possess this ancient sense, but not keenly, and make little conscious use of it. Hunters teach themselves to compensate for the superior sense of smell of many prey animals by approaching them from downwind or by disguising the human scent with others. Given the powerful olfactory sense of so many animals, it is possible they have aesthetic responses to stimuli people do not detect. An observer of coatimundis in Arizona reported that these animals frequently sit up or lean back and sniff the air intently, presumably gathering information. He commented that one old female, the Witch, would also sometimes arise during one of the group's intervals of relaxation on a cliff ledge, go to the edge and sit for five minutes or so, sniffing calmly, slowly, and deeply. The thought that she might be appreciating and not just assessing the world around her occurred to observers, who could not resist comparing her to a concert-goer or gallery visitor.

The frequent allegation that birds have no sense of smell is simply wrong. The olfactory bulbs in a bird's brain give them this sense. Its acuity varies widely: parrots and warblers seem to have poor senses of smell, while those of albatrosses and kiwis are excellent. Not surprisingly, some vultures have been found to have an excellent sense of smell, which they use to locate carrion. It might be pleasing or merely interesting to their aesthetic sense.

Many snakes have special heat-detecting organs. An increasing number of ocean creatures are found to have electromagnetic navigation senses. There are some senses, such as the ability to detect magnetic fields, that science has only recently become aware of; there may be more. Any sense may be subject to preference and any field of preferred and less preferred points may be perceived as beautiful or ugly. Some animals may be admiring or even creating subsonic, infrared, or electromagnetic beauty.

One scientist who gathered data about the visual preferences of immature male rhesus monkeys in captivity found that they liked short-wavelength light. They preferred orange to red, yellow to orange, green to yellow, and they liked blue better still. They were more interested in pictures of other animals than they were in pictures of monkeys, but preferred pictures of monkeys to pictures of people. They would rather look at flowers than at a Mondrian painting (Composition, 1920), and pictures of bananas interested them least of all. They would rather watch a continuous cartoon than a cartoon film loop, but they preferred the loop to watching still pictures. They preferred their films to be in focus, and the worse the focus was on a film, the less they cared to watch it. These preferences were categorized as "interest" and "pleasure." The preference that seems the most likely to be purely aesthetic is the one for colors, since there is no reason to suppose that a plain blue wall is more interesting than a plain yellow one. Nor are rhesus alone in their taste for blue. The German animal behaviorist Bernhard Rensch also examined color and pattern preferences in primates and in other animals. The apes and monkeys generally preferred regular to irregular and symmetric to asymmetric designs. Nor were their tastes immutable: retested after an interval, some of them made different choices. When Rensch tested crows and jackdaws, they, too, selected regular designs, but fish seemed more attracted by irregular designs.

Animals often select mates on the basis of their displays or songs. Sometimes the criteria they use can be simply quantified: they select the biggest or the loudest or the plumpest mate. Female widowbirds are attracted to male widowbirds with long tails, so when an ornithologist glued extra tail feathers onto the end of the tails of some males, those males became more popular. They tend to favor animals with a symmetric appearance, also a preference of some humans. Sometimes, however, more subtle aesthetic choices seem to be involved.

The beautiful and unusual bowerbirds and birds of paradise of New Guinea are favored subjects of ornithological study. The various species of bowerbirds do not form pairs. Instead, the female visits the display sites, or bowers, of various males and the male performs courtship displays, which may or may not induce the female to mate with him. Some male bowerbirds—generally those with the plainest plumage—construct very elaborate bowers resembling alleys, tunnels, maypoles, yards, or tepees. They embellish these with colored objects such as flowers, fruits, insect parts, or human artifacts; and they may even paint parts of the bower with charcoal and crushed berries, using a bark "brush."

Different populations of a bowerbird species prefer different colors when choosing curios to decorate their bowers. They often visit one another's bowers and steal decorations. When selecting adornments, the satin bowerbird—a blue-eyed species—likes blue items. When the male bowerbird decorates his bower, and when (and if) the female bowerbird mates with the male whose bower she likes best, they seem to be exhibiting taste. When interfering biologists steal adornments from certain bowers, females favor them less and those males are able to mate less often. Perhaps for various reasons correlated with fitness, blue-loving bowerbirds have a reproductive competitive advantage. The evolutionary approach would suggest that the ultimate cause of their artistic preferences may be to enable the male to show how fit he is, how much time he has to devote to collecting decorations and defending them from theft, and for the female to assess this. The proximate cause, however, is unlikely to be anything like this. It is unlikely that the female estimates how many bird-hours went into a bower, and whether that indicates good genes. The male does not decide to decorate with blue objects because, let us say, they are rare, and he knows that will indicate to females that he ranges over a wide area due to his good genes. It is more likely—and it is a more parsimonious theory—that both female and male bowerbirds like the look of blue.

Naturalist Bruce Beehler observed the bowers of the streaked bowerbird, birds who are not born especially beautiful, but must create their own beauty, hence their bowers. These resemble tepees with center poles, and on the base of the pole the bird constructs a tidy wall of moss adorned with colored objects. Each type of decoration is on a different section of the wall and the effect is "quite artistic and very beautiful." Beehler notes: "Some biologists believe that the remarkable construction of the male bowerbirds is evidence of an aesthetic sense. Others prefer to believe that this spin-off of mating behavior is the product of the remarkable sexual competition among males to mate with females—a process that Charles Darwin named 'sexual selection.' " These two explanations are not opposed, but compatible. Yet something important is suggested by the comment that some biologists "prefer" to believe in competition. This is the question of what people *like* to believe or think they ought to believe about animals.

The same issue arises with respect to birds of paradise, famous for the gorgeous plumage of the males. At a display site of the lesser bird of paradise, many males gather. All of the females who visit may mate with just one male, and biologists have wondered why. Beehler writes:

> Some researchers believe that it is the product of acute female discrimination and that the females are choosing the "prettiest" or "sexiest" male. I tend to believe that it is caused, in part, by a despotic control of the lek [communal breeding ground] mating hierarchy by the dominant bird. The alpha male, by periodic physical aggression and continual psychological intimidation, is able to keep control over the subordinate males in the lek. The females are able to perceive this hierarchy in the lek, just as humans can make the same perception about dominance and subordination in a social situation. Females will naturally tend to mate with the alpha male, because his genetic material will most likely give the female the best chance of producing offspring with his qualities—the qualities that may help her male offspring dominate a lek of the next generation.

This analysis, with its focus on male aggression and its rewards, leaves unexplained the huge golden tail feathers that the males vibrate in their displays, found so beautiful by humans that the very existence of some species has been threatened by hunting for export. If females can perceive something as complex as hierarchy, and mate according to it, why is it impossible for them to enjoy and be attracted to shimmering gold?

New Guinea human tribes have differing styles of ritual costume, which almost invariably include feathers from various species of birds of paradise, usually on the men's headdresses. Bowerbirds often take human artifacts such as brightly colored candy wrappers, cartridge cases, or shiny keys to decorate their bowers. One can imagine birds creeping near human habitations to steal colored objects and humans creeping near bird habitations to steal colored feathers. When we do it, it is for art. When they do it, it is for competition. Both may be true. What is disturbing and irrational is the decision to explain human behavior in spiritual terms of a sense of beauty, and animal behavior in mechanistic terms of demonstrating fitness. The object, yet again, seems to be to define humans as higher and unique.

Artistic Creation

The issue of artistic creation in animals is fascinating, but little work has been done in this area. This is one of the many activities that has been said to mark a boundary between humans and animals. Various apes, particularly chimpanzees, have drawn or painted in captivity, as have capuchin monkeys. Alpha, a chimp at the Yerkes laboratory, loved to draw and would beg visitors for paper and pencil—in preference to food—and then sit in a corner and draw. Once, lacking paper, she tried to draw on a dead leaf. By giving her paper that had geometric designs on it, it was found that her drawing was influenced by what was already on the paper. She filled in some figures, scribbled in the missing parts of others, such as a circle with a wedge cut out, and added marks that "balanced" other figures. Her drawings were promptly taken from her, because after she had drawn on both sides of a sheet of paper, she would put it in her mouth.

Following this work, Desmond Morris had no difficulty in persuading another chimpanzee, Congo, to draw and paint. Interrupted before he had finished a picture, Congo screamed with rage until allowed to complete it. Congo also changed his drawing depending on what was already on the paper. His favorite design or theme was a fan of radiating lines, which he made in various ways, not by a single, stereotyped technique. Gorillas such as Koko and Michael have also made many drawings. In no case have apes produced incontrovertibly representational drawings. One chimp, Moja, produced an unusually simple drawing with parallel horizontal curves, and signed that it was a bird. Asked to draw a berry, she produced a compact drawing in one corner of her paper. Either drawing is plausible but not irrefutable as a representation, but then many humans create and value unrepresentational art.

In a later experiment Moja and Washoe were asked to draw

Fig 10-1. Alpha, an 18-year-old caged chimpanzee at the Yerkes laboratory, for years regularly begged humans for pencil and paper so that she could draw. Once, when she could not get paper, she drew on a dead leaf. If another chimpanzee was in the cage when she was drawing, Alpha would shoulder it away or turn into a corner. She made this drawing in red and blue crayon on 8″ × 11″ white paper, over a period of three minutes. Alpha was never rewarded for drawing and would ignore food if she saw a chance to get pencil and paper. [From the *Journal of Comparative and Physiological Psychology*, published by the American Psychological Association]

such items as a basketball, a boot, a banana, an apple, a cup, and a brush, either from the actual objects or from color slides. At later sessions they were asked to draw the same items and the drawings were examined for consistency. Drawings of the boot were inconsistent, but those of the cup and brush showed similarities. None of these were drawings that a human would be likely to identify as cups or brushes: requests to draw the cup always produced a centrally placed solid vertical fan of strokes, whereas the brush was denoted by vertical strokes crossing horizontal strokes. Drawings of flowers included radial patterns and drawings of birds always included "a pointed action," whether denoting a beak, the movement of flight, or something else is unknown. "The one that threw us with Moja was the basketball, in that it just was a scribble across the page," Roger Fouts said. But after Moja drew it in the same way—vertical zigzags on the lower part of the paper—at intervals six weeks apart, researchers realized it might represent not the appearance of the ball, but its motion. Small children sometimes produce kinesthetic drawings of this kind.

While the boredom of captivity may be a motivating factor for all such animals, they appear to find the act of drawing or painting

Fig. 10-2A. This painting (watercolor on paper) was produced by the young chimpanzee Congo. Desmond Morris, who encouraged Congo to draw and paint, notes that fan patterns like this were one of Congo's favorite themes. The lines were made by moving the brush toward his body.

Fig. 10-2B. This fan pattern was made by Congo at the same session as the previous painting, but, astonishingly, was made in a completely different manner: Congo moved the brush *away* from his body, grunting softly as he paused to study the lines. The fact that Congo could make similar patterns by differing methods shows that he was not simply repeating stereotyped motions.

rewarding in itself. (As if to once again remind us that animals remain individuals, some captive chimps absolutely refuse to draw or paint.) In 1980 a young Indian elephant named Siri (the same who rubbed fruit into her hay) was assigned a new keeper, David Gucwa. Noticing Siri making scratches on the floor of her enclosure with a pebble—and then "fingering" them with the tip of her trunk—Gucwa began providing Siri with a pencil and drawing pad (which he held in his lap). She responded by producing dozens of drawings. All could be classified as either "abstract" or scribbles, but all are confined to the boundaries of the paper and, to many observers, seem lyrical, energetic, and beautiful. She was never rewarded with food for drawing, although she might have found Gucwa's attention a reward in itself.

Gucwa and journalist James Ehmann sent copies of Siri's drawings to scientists, most of whom declined to comment, and to artists, many of whom were enthusiastic. Artists Elaine and Willem de Kooning, in particular, looked at the drawings before reading the covering letter, and were struck by their "flair and decisiveness and originality." Learning the identity of the maker, Willem de Kooning remarked, "That's a damned talented elephant." (Given her circumstances, Siri could hardly be derivative.) Copies of her drawings were shown to other zookeepers, who said that was nothing new: *their* elephants scribbled on the ground with sticks or stones all the time. Why, then, had nobody written about it before?

Siri's opportunity to draw on paper ended after two years due to differences between Gucwa and the zoo director, and to her transfer to another zoo during renovations. She was never given paper with predrawn designs on it to see how that would affect her drawing, but on several occasions she made two drawings on one sheet of paper, and apparently placed the second with reference to the first. We do not know whether Siri ever disliked any of her work—did she ever tear anything up? Gucwa always removed her drawings promptly so that she wouldn't smear them with her damp trunk when she "fingered" them. If other elephants also like to draw, how would they react to one another's drawings?

Other elephants have made images on paper or canvas, but perhaps none with so little direction as Siri. Carol, an Indian ele-

Fig. 10-3A. Drawing done by Siri, an Indian elephant, in pencil on 9″ × 12″ paper. She drew on a sketch pad that handler David Gucwa held in his lap. Gucwa titled this "I Remember Swans," and included it with others sent to artists Elaine and Willem de Kooning. Before learning that an animal drew them, they admired the flair and originality of the sketches and, when he learned the artist's identity, Willem de Kooning exclaimed, "That's a damned talented elephant."

phant at the San Diego Zoo, was taught to paint as a visitor attraction, and is given commands by her trainer, who tells her when to pick up the brush, supplies her with colors, rotates the canvas so the brush strokes go in different directions, and rewards her with apples. But just as the existence of paint-by-numbers kits does not invalidate the existence of artistic originality, Carol's dutiful paintings do not negate Siri's apparently spontaneous urge to draw.

More recently, Ruby, an Asian elephant in the zoo at Phoenix, Arizona, has been encouraged to paint. Selected because she was the most active—but not the only—doodler among the elephants, Ruby loves to paint and becomes excited when she merely hears the word *paint*. Blue and red are by far her favorite colors. Biologist Douglas Chadwick reports that she may select other colors that correspond to an unfamiliar object nearby, so that if an orange truck is parked in her view, she may pick orange paint. "A zoo

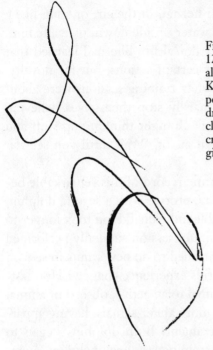

Fig. 10-3B. This drawing (pencil on 9″ × 12″ paper), titled "Iris" by Gucwa, was also among those praised by the de Koonings. Before she was given the opportunity to use pencil and paper, Siri drew in the dust on the floor of her enclosure, scratching designs on the concrete with sticks and pebbles. She was given no rewards for drawing.

visitor was once taken ill while watching Ruby paint, and paramedics were called to the scene. They wore blue suits. It might have been a coincidence that after they left, Ruby painted a blue blob surrounded by a swirl of red." She still doodles in the dirt in her enclosure. A handler thinks the African elephants who share the yard with Ruby are jealous of the attention she gets, because they have begun making highly visible drawings on the walls, using the ends of logs.

A different kind of creativity was seen in rough-toothed dolphins. Trainer Karen Pryor had run out of stunts for the dolphins to perform in shows and decided to reward one of them, Malia, only for performing new behaviors. The trainers waited until Malia did something new, then rewarded her by blowing a whistle and tossing her a fish. Malia quickly learned what was being reinforced that day—tail-slapping, backward tail-walking—and performed that action. In a few weeks they had run out of new behaviors to reinforce. After several frustrating days, Malia suddenly began performing a dramatic array of wholly new activities, some quite complex. She swam on her back with her tail in the air, or spun like a corkscrew, or jumped out of the water upside down, or made lines on the floor of the tank with her dorsal fin. She had learned that her trainers were not looking for certain actions, but for novelty. Sometimes she was very excited when training sessions were about to start, and her trainers could hardly stop themselves believing that Malia "sat in her holding tank all night thinking up stuff and rushed into the first show with an air of, 'Wait until you see *this* one!' "

Determined to make a scientific record of this remarkable behavior, the trainers filmed the same process with a second dolphin, Hou. A less optimistic and excitable individual, Hou took longer to catch on, but at the sixteenth training session suddenly performed a burst of novel activities and continued to do new stunts in session after session. Pryor reports that this experience changed Hou lastingly, "from a docile, inactive animal to an active, observant animal full of initiative." Hou also gave more anger signals, having apparently acquired an artistic temperament. Both dolphins began to produce novel behaviors outside training sessions, including opening gates between tanks, leaping over gates, and jumping out of the water and sliding around on the concrete in order to rap trainers on the ankles. While some thought these results showed how intelligent dolphins are, Pryor disagreed: she replicated the experiment with pigeons and ended up with birds that spontaneously lay on their backs, stood with both feet on one wing, or flew into the air two inches and hung there. Such creativity is unexpected in a species usually deemed intellectually unimpressive, but may simply show that, given a chance to be creative, they are smarter than we thought they were.

Culture and the Concept of Beauty

Human emotions occur in the context of culture, which is not to say that they exist because of culture. Our sense of what is beautiful is significantly influenced by culture. Different human cultures teach somewhat different ideas of what is beautiful. For example, music that delights one group of people may strike members of another culture—or subculture—as discordant, depressing, or ugly. There is more to human perception of beauty than culture, but exactly how much more is unclear. Thus the idea that animals perceive and create beauty leads to the question of whether their aesthetic perception is in some way related to culture.

Culture and its transmission include much that is cognitive, which is not the focus of this book, yet it should be noted that there are many examples of animal culture. In Japanese monkey troops, different troop traditions have been noticed: some troops eat shellfish and some do not; some eat the seed of the muku fruit and some throw it away; some baby-sit for each other and some do not. The most famous example of cultural transmission comes from the story of Imo, the "monkey genius" who invented several feeding techniques, such as throwing handfuls of sandy grain into water so that the sand sank and the grain could be scooped off the surface of the water and eaten. Imo's methods were gradually copied by more and more of the monkeys in her troop until they all practiced them.

Elizabeth Marshall Thomas has written of prides of lions in the Kalahari Desert in southern Africa that had a tradition (later lost) of coexistence with humans. In the 1950s, when Thomas first lived in the Kalahari, lions treated humans—mostly Juwa and Gikwe Bushmen—with resentful respect. They reluctantly allowed humans to drive them away from kills and did not attack humans. In the 1960s the Bushmen were evicted from the area. When Thomas returned in the 1980s, she found that the lions behaved differently. These lions, who no longer lived among a human population, had lost their cultural traditions toward them and seemed all too willing to examine the idea of humans as prey. Thomas also writes that leopards in different areas used different hunting methods for human prey. In what may be described as a cultural shift, a group of olive baboons in Kenya was observed to switch from occasional meat-eating by adult males to a custom of hunting and eating prey by males, females, and juveniles. A curious example from captivity is supplied by the chimpanzees at Arnhem Zoo. A dominant male injured his hand in a fight and had to support himself with his wrist while walking, whereupon the young chimpanzees began hobbling on their wrists too.

In the University of Washington group of signing chimpanzees where Loulis grew up, the adult chimpanzees had been raised by Allen and Beatrix Gardner, whom Loulis had never met. One day the Gardners came to visit, an event that was not announced to the chimps. None of the chimps had seen the Gardners for at least a year, and Washoe had not seen them for eleven years. The Gardners walked in. When the chimps saw them they simply sat and stared—highly unusual behavior. They did not give the usual friendly greetings—signs, touches, hugs—given to people they know, nor did they display as they do at strangers, but gazed as if dumbfounded. Except Loulis, who began to display at these unfamiliar people, erecting his hair, standing and swaying, banging on walls. Instantly, Washoe and Dar, who had been sitting on either side of him, seized him. Dar clapped a hand over Loulis's mouth. Washoe took his arm and made him sit down, which he did with an astonished expression. Such treatment was new to him. After a moment Washoe went to the Gardners, signed their names, led Beatrix Gardner into another room, and began to play a game with her that they had played together in Washoe's childhood. In this instance, a piece of cultural information was transmitted to Loulis. The chimpanzees' signing abilities did not extend to making the statement, "These are people we treat with affection and respect," yet Loulis surely was given this message.

Culture has been described as yet another factor that makes humans unique. Examples of cultural transmission of behavior like those above are dismissed as "just interesting oddities." They may be more than that. Cultural transmission may be more widespread among animals than is commonly believed. It has not been argued that any animal species lives in a culture approaching the complexity of any human culture. However, the contention that animals cannot possess a sense of beauty because aesthetics are culture-bound is wholly unproven and there is much evidence to support the contrary.

Just as any two people do not find all the same things beautiful, a person and an animal might disagree. But to dismiss the possibility that an animal can sense beauty at all is narrow-minded. Discussing birdsong, Joseph Wood Krutch wrote:

> Suppose you had heard at the opera some justly famous prima donna singing "Voi che sapete." . . . You have assumed that [she] genuinely loves music and experiences some emotion related to that which Mozart's aria expresses. . . . But a scientist of another kind—an economist—comes along and says, "I have studied the evidence. I find that . . . the performer really sings for so many thousand dollars per week. In fact she won't sing in public unless she is paid quite a large sum. . . . You may sing in the bath because you are happy and you like to do it. But so far, at least, as professional singers are concerned, they sing for nothing but money." The fallacy—and it is the fallacy in an appalling number of psychological, sociological, and economic "interpretations" of human behavior—is, of course, the fallacy of the "nothing but" . . . there is nothing in human experience or knowledge to make it seem unlikely that the cardinal announcing from the branch of a tree his claim on a certain territory is not also terribly glad to be doing it, very joyous in his realization of his own vigor and artistry. . . . Whoever listens to a bird song and says, "I do not believe there is any joy in it," has not proved anything about birds. But he has revealed a good deal about himself.

Researchers studying elephants in Kenya camp in the middle of the east African bush. Sometimes at night the people sing and play guitars and the elephants draw near to listen. Perhaps they are merely curious, but perhaps they take pleasure in the music. Human curiosity should allow us to ask whether elephants find beauty in music, just as the human sense of beauty allows us to appreciate the image of these large beasts moving slowly through the darkness listening to the songs.

The Religious Impulse, Justice, and the Inexpressible

While nonhuman animals show clear evidence of emotional lives, they should not be considered emotionally identical to humans. This would be true anthropomorphic error, inaccurate projection. We are all beings together, but we are not the same—neither higher nor lower, just distinct.

Much past discussion of animals and emotions has been directed primarily at arguing that certain emotions are the property of humanity alone. To show that, on the contrary, some animals feel some emotions, still means that the question must be asked anew in particular instances. If hippos can feel compassion, that does not mean that a particular hippo feels compassion at a particular time. Similarly, if buffalo are capable of love, this does not necessarily mean that they can feel shame. So it is still possible that there are emotions people feel that no other animals feel. Leaving aside the unimpressive history of human attempts to claim uniqueness, it should be conceded that many species do have attributes that no others have. Pelicans have unusual bills, elephants have trunks, platypuses have poison spurs; perhaps human beings can claim religious awe.

Religion and the Soul

People have immortal souls and animals do not, according to much of Western religion. Animal lovers resist this, citing animals' virtues and affirming that they must have souls; heaven would be a paltry place if there were no dogs in it. The question of who has a soul and who doesn't is far more problematic than the parallel question of emotions. Science is of no help. Yet the theological view may point to a difference between the emotional lives of humans and those of other animals. Animals do not seem to need to believe in higher powers. Animals have not been observed to have religious practices, while people do.

Some traditional tribes in Madagascar say that when sifaka lemurs lie on high branches in the morning, facing the sun, with their eyes closed, they are worshiping the sun. Some say that the sifakas are incarnations of their own sun-worshiping ancestors. Primatologist Alison Jolly commented, "It is difficult to watch a sunning lemur without being anthropomorphic, but to Western eyes it seems less like religious fervor than like our indolent cult of Sunday at the beach." There is no reason to suppose that sifakas themselves are lemuromorphic, that they invest the sun with creature qualities (though we cannot prove that they do not) and worship those qualities. The explanation that they enjoy the warmth seems sufficient to explain their behavior, but the traditional explanation of the Malagasy people has the advantage of poetry.

Like art, religion is not an emotionless affair of pure intellect. Awe, faith, righteousness, abasement, worship, seeking salvation— all have emotional components. Some theorists describe awe as a form of shame. Are the religious emotions ones that animals simply do not feel? Or are they emotions that exist in other parts of life that in humans may be focused on the religious impulse? Animal emotions may shed some light. Elizabeth Marshall Thomas compares the behavior of a person humbly kneeling to pray and a dog showing its belly—demonstrating submission—to a person. Her husband's dog, she notes, ritually shows his belly first thing in the morning, like a morning devotion. In the end, Thomas concludes that the parallel is not exact, that dogs probably do not think of humans as gods, yet, "as we need God more than he needs us, dogs need us more than we need them, and they know it." Further study of such dynamics could shed considerable comparative light on human religious rites.

Morality and a Sense of Justice

The sense of justice has been called uniquely human. Has it an emotional component? The human sense of justice is accompanied by many emotions: anger and outrage at injustice, desire for revenge, and compassion. The stories of crow parliaments that hold trials and pass judgment on their members are fantasy, but less organized manifestations of what may be a sense of justice exist in many stories of righteously indignant chimps, for example. Nim Chimpsky learned when to expect praise and when to expect censure, and accepted these artificial standards. If he broke a toy, punishment did not surprise him, and he accepted it. But if one of his teachers punished him for something the others ignored, or if one failed to praise him for something the others rewarded, Nim became sulky.

Perhaps Nim was merely upset because his world lacked predictability, because his settled expectations were violated. But this is a big part of legal justice among humans. The chimpanzees in the Arnhem colony seem to react to a sense of unjust treatment of others. In one instance the chimpanzee Puist "kidnapped" a one-year-old infant from his mother, and carried him up a tree, where he screamed with fright. After the mother recovered her child she attacked Puist, although Puist was larger and more dominant. The male Yeroen rushed up to them and stopped the fight by seizing Puist and flinging her away. This was unusual, because Yeroen and Puist were allies, and on all other occasions Yeroen had intervened on Puist's side. Frans de Waal concludes that Yeroen agreed with the mother chimp that she had cause for complaint.

In another incident, Puist appeared to be aggrieved on her own behalf, backing up Luit in a dispute with a large male. The male made a threatening display at Puist, who stretched out her hand in appeal to Luit. Luit did not respond, and Puist instantly rushed at him, barking and even hitting him, apparently because he had violated the tradition of supporting one's supporters. This kind of solidarity is part of many human notions of fairness.

Elizabeth Marshall Thomas gives an instance that may illustrate a sense of justice or perhaps a sense of propriety that underlies rectitude. Her husky, Maria, one day discovered that by rushing around a cage full of mice and parakeets, lunging at the occupants, she could send them into a hysterical frenzy. The pug Bingo ran in, slammed his body into the much bigger Maria, and when she lunged at the cage again, uttered a loud bark and slammed against her once more. Maria left the room. Thomas was surprised because Bingo was smitten with Maria and usually did not oppose her in any way. Whether Bingo was motivated by compassion for the mice and parakeets, by vicarious ownership of the mice and parakeets, or by disapproval of Maria's obstreperousness, it seems undeniable that he wanted to stop her aggression and make her behave better toward the other animals.

Observers of wild coatimundis in Arizona suggest that they have a system of entitlement expressed through a variety of cub squeals. When cuffed by an older animal for lagging behind the troop, a cub would crouch submissively and utter the "don't beat me" squeal, which seemed to indicate no resistance. On several occasions, when a subadult animal committed the unusual act of trying to take food from a cub and cuffed it, the cub would utter a different squeal—and an adult female would come and drive the subadult away, apparently enforcing a tradition of tolerance toward cubs. This may be merely different cub feelings being expressed in different situations of threat, but it is telling that there is a difference. Enforcing and cushioning hierarchy also plays a part in human justice systems.

The Narrative Urge

The desire to tell stories is another characteristic of humanity. People like to relate events, gossip, analyze. Humans talk to animals and to ourselves. Does language itself create the narrative urge, or would human beings have this need even without human language?

Animals who have been taught sign language have been said to display little wish to narrate. Herbert Terrace, who arranged for the chimpanzee Nim Chimpsky to be taught sign, said that most of Nim's utterances were imitations or fragments of things his teachers had just signed, and argued that the same is true of other signing apes. He also argued that a large part of ape signing consists of requests for food, toys, and affectionate gestures like tickling and hugging. This, and the scarcity of spontaneous verbal communication, would seem to indicate little narrative urge. On rare occasions, Nim would sign unasked the names of things he saw. Often he spontaneously signed the names of things he recognized in pictures when leafing through books and magazines. Perhaps these are rudiments of a narrative urge, awaiting encouragement and opportunity.

Nim's language lessons (as in most such experiments) were structured to offer him the chance to earn food and other rewards, so it is not surprising that he made many such requests. It is also worth noting that Nim's early teachers were not fluent signers. Most of them could only improvise a few sentences on any given topic—they could not tell Nim a story, relate the events of their day, or pass on interesting gossip. Nim began learning sign at five months, but he did not get a fluent teacher (and then not for long) until he was three and a half years old. This is not unusual. Signing apes have been raised and taught largely by humans who use rather rudimentary sign language. None have been raised in an environment of fluent sign. Suppose a child were raised by people who spoke halting, recently learned pidgin. Suppose also that the child had no playmates or classmates with whom to speak. Such a child might fall linguistically behind children whose parents freely and literately spoke to each other and to others, as well as to the children. A child who never witnessed a story being told might not tell stories, but this would not demonstrate the limits of human storytelling ability.

Terrace mentions that when Nim met fluent signers, he was

transfixed. He would stare spellbound for up to fifteen minutes (a long time for a young chimpanzee) as they conversed. In contrast, spoken language interested him for only a few seconds. Terrace notes that when, at three and a half, Nim finally got a teacher fluent in sign—apparently his fifty-fourth teacher—he was already passing into adolescence. Terrace thinks Nim's signing might have progressed more quickly had he had more exposure to fluent signers at an early age. Washoe, the first chimpanzee to be taught sign language, adopted a son, Loulis, who has learned sign language not from human tutoring but from Washoe and the other chimps in her colony. Yet Washoe herself did not learn sign from fluent signers. It is possible that apes have not yet been adequately challenged to acquire sign fluency. If so, there has not yet been a full test of their possible desire to narrate.

Signing chimpanzees other than Nim have been seen to sign to a considerable extent in the rudiments of a narrative mode. They sign to each other even when no humans are around (as revealed in remote videotaping) and, like humans, "talk" to themselves. Washoe has been filmed perched in a tree, hiding from her human companions, and signing "quiet" to herself. They may describe their own activity to themselves, signing "me up" and then running up a wall. They have even been seen to use imaginative speech when playing alone. Moja, who knows the word "purse" perfectly well, once put a purse on her foot and walked around signing "That's a shoe." Thus begins the rudiments of metaphor.

There is a sense in which bees are entirely narrative, letting other hive members know where the best flowers are and how to get there. The most revolutionary discovery of Karl von Frisch concerned the symbolic communication employed by honeybees: a bee who has found flowers performs a dance when she returns to the hive, which tells other bees how far away the food is and in which direction. Donald Griffin notes, "In the scientific climate of opinion prevailing forty years ago it was shocking and incredible to be told that a mere insect could communicate to its companions the direction, distance and desirability of something far away." But this is exactly what they did.

Chimpanzees Sherman and Austin were taught to communicate by lighting symbols on a board. Researcher Sue Savage-Rumbaugh notes that they can use these symbols to make spontaneous comments about their impending actions and about events going on around them, but that they do so relatively seldom. "Their behavior suggests that it is difficult for them to understand that others do not have access to the same information that they do. In the various paradigms used to encourage communication between them, it was always necessary for them to experience the roles of speaker and listener a number of times before their behavior, as speaker, suggested that they knew that they had information which the listener did not have," Rumbaugh writes. While Sherman and Austin learned about the possible ignorance of the listener in individual situations, and do not seem to have generalized their observations, it is not impossible that they might do so. They were taught to share food with one another in a most unchimpanzeelike manner and came to enjoy this greatly, although it was apparently not easy for them to learn. Perhaps they could learn narration in a similar way.

It is possible that great apes will never show more language ability than they have so far, that they have reached their linguistic limits. It may well be that the urge to confide, to boast, to retell, and to mythologize will remain a human trait, but too little is known to be confident of this. If, instead of devising laboratory conditions under which apes can learn to communicate with humans in our language or some variant of it, humans were to go quietly into the forest and listen to what is already being communicated, more would be learned. Some of the most vocal animals are very little understood. Some species of whale are clamorous, incessantly uttering a wide variety of squeaks, grunts, trills, bellows, chirps, groans, yaps, and whistles as well as echolocatory clicks and pings. Perhaps these mean only "Here I am. Where are you?" An alternative view is given by Jim Nollman, who commented on the finding of whale scientist Dr. Roger Payne that humpback whales may repeat their entire song from one year to the next, "with slight but clearly discernible differences. Here was a very clear instance of an oral tradition. It implies that humpbacks possess at least the rudiments of learned culture." Perhaps they are telling the history of the species.

The search for emotions that humans feel and other animals do not is an old one. To reverse this and look for emotions that animals have and we do not violates the usual assumption that humans are the perfected endpoint of evolution and the luckiest recipients of nature's gifts. But it is impossible not to acknowledge the many things some animals have that we do not. Some we are proud of not having: tails, fur, horns. Some we shrug off: a keen sense of smell. Others we envy: wings.

Exclusively Animal Emotions

Some animals have senses humans do not possess, capacities only recently discovered. Other animal senses may remain to be discovered. By extension, could there be feelings animals have that humans do not, and if so, how would we know? It will take scientific humility and philosophical creativity to provide even the beginning of an answer.

A mother lion observed by George Schaller had left her three small cubs under a fallen tree. While she was away, two lions from another pride killed the cubs. One male ate part of one of the cubs. The second carried a cub away, holding it as he would a food item, not as a cub. He stopped from time to time to lick it and later nestled it between his paws. Ten hours later, he still had not eaten it. When the mother returned and found what had happened, she sniffed the last dead cub, licked it, and then sat down and ate it, except for the head and front paws.

This mother lion was acting like a lion, not like a person. But in understanding what lions do, what she felt is part of the picture. Maybe she felt closer to her dead offspring when it was part of her body once again. Maybe she hates waste, or cleans up all messes her cubs make, as part of her love. Maybe this is a lion funeral rite. Or maybe it is something only a lion can feel.

Elephants display a behavior called "mating pandemonium." When a female elephant in estrus mates, she utters a loud call, in registers too low for the human ear. When they hear the call, her relatives come racing to the scene, trumpeting loudly, appearing agitated or excited, and pandemonium ensues. Other male elephants may also be attracted. Unrelated groups ignore the call or leave the area. As observer Joyce Poole has remarked, "Biologically, you could say that mating pandemonium serves to attract still more males to the . . . female, increasing the chances that a still more dominant bull will come and drive off the one guarding her and end up being the one to actually fertilize her. I happen to think mating pandemonium is more than that, but whether it has to do with social territories, some type of emotional support for the female in heat, or something else altogether, I couldn't say." What are the emotions of the female's relatives, the ones creating the pandemonium? The answer is unclear. They might be feeling a mix of many emotions, known and unknown.

After thirty years of working with chimpanzees, Roger Fouts doubts that they have emotions that humans do not also have. Indeed, if new and unknown emotions were to be discovered, they would most likely be found in animals less like us than the great apes. One spring evening, George Schaller watched a female wild giant panda in China, Zhen-Zhen, eat and then—although she saw him watching her—lean back on some bamboo, uttering "bleating honks," and fall asleep. Her seeming indifference surprised Schaller:

> On meeting a gorilla or a tiger, I can sense the relationship that binds us by the emotions they express, for curiosity, friendliness, annoyance, apprehension, anger, fear are all revealed by face and body. In contrast, Zhen and I are together, yet hopelessly separated by an immense space. Her feelings remain impenetrable, her behavior inscrutable. Intellectual insights enrich emotional experiences. But with Zhen I am in danger of coming away empty-handed from a mountain of treasure.

This is not to say that pandas are unknowable. Schaller believes that he could learn to understand. "To comprehend her, I would need to transform myself into a panda, unconscious of myself, concentrating on her actions and spirit for many years, until finally I might gain fresh insights." He fears pandas may not last long enough as a species for humans to come to understand them.

Unconscious Emotions

Even if animals have emotions, some argue, they do not feel them as human beings do, because they cannot be aware of them, bring them into consciousness and express them to themselves. Perhaps an elephant can be sad, the argument goes, but if it cannot say, even to itself, "I am sad," then it cannot be sad in the same way that a person is—who can describe sadness, predict sadness, lose an argument with sadness. If this is true, then it is language that has given humans their tremendous attachment and vulnerability to their feelings. It is rash, at this point in knowledge, to be at all certain that an emotion that cannot be expressed in language, far less in language we can recognize, cannot be felt as keenly.

Humans believe they suffer from emotions they are not consciously aware of and that are unarticulated. This does not mean that these emotions have no significance or cannot really be felt. It could equally well be argued that language sets emotion at a distance, that the very act of saying "I am sad," with all the connotations that the words have, pushes the feeling away a little, perhaps making it less searing and less personal. Herbert Terrace describes what might be an actual example of language pushing feeling away in an animal:

> Certain usages of Nim's signs were quite unexpected. At least two of them (*bite* and *angry*) appeared to function as substitutes for the physical expression of those actions and emotions. Nim learned the signs *bite* and *angry* from a picture book showing Zero Mostel biting a hand and exhibiting an angry face. During September 1976, Amy began what she thought would be a normal transfer to Laura. For some reason, Nim didn't want to leave Amy and tried to drive Laura

away. When Laura persisted in trying to pick him up, Nim acted as if he was about to bite. His mouth was pulled back over his bare teeth, and he approached Laura with his hair raised. Instead of biting, however, he repeatedly made the *bite* sign near her face with a fierce expression on his face. After making this sign, he appeared to relax and showed no further interest in attacking Laura. A few minutes later he transferred to Laura without any sign of aggression. On other occasions, Nim was observed to sign both *bite* and *angry* as a warning.

To the extent that language pushes feelings away, the world of emotion, a world from which humans sometimes feel estranged, may be one some animals live in *more* fully.

Emotional Intensity

Whether some animals feel emotions less or more intensely than humans may depend on which emotion is involved. Animals doubtless feel pity for one another, sometimes even passing beyond the species barrier, but it seems unlikely, though not impossible, that they experience it as elaborately or as intensely as humans do. For example, it is doubtful that the dolphins care as much about humans slaughtering one another as some humans care about the slaughter of dolphins by other humans. But this may only be because they do not have the same access to information that humans do. Perhaps they know and have rules of noninterference in human matters. Perhaps they truly are indifferent, or take a longer view.

There are some emotions, on the other hand, that humans may experience less intensely than some animals. Many people have had the feeling, for example, that some animals seem more capable of joy. One of the explanations for the popularity of watching and listening to birds is the pleasure of hearing birdsongs, which seem to them joyful. As Julian Huxley, describing the courtship of herons winding their long necks together, wrote: "Of this I can only say that it seemed to bring such a pitch of emotion that I could have wished to be a heron that I might experience it."

The intensity of emotion in other animals has been a perennial source of human envy. Joseph Wood Krutch writes: "It is difficult to see how one can deny that the dog, apparently beside himself at the prospect of a walk with his master, is experiencing a joy the intensity of which it is beyond our power to imagine much less to share. In the same way his dejection can at least appear to be no less bottomless. Perhaps the kind of thought of which we are capable dims both at the same time that it makes us less victims of either. Was any man, one wonders, ever as dejected as a lost dog? Perhaps certain of the animals can be both more joyful and more utterly desolate than any man ever was."

To examine questions like these it is vital to treat animals as members of their own species. Treating them as either machines or people denigrates them. Acknowledgment of their emotional lives is the first step; understanding that their emotional lives are their own and not ours is the second. At the same time, if humans have no peers as cognitive beings and creatures of elaborate cultures, as emotional beings we are anything but alone. Is there a reason why we should try to comprehend the world of animal emotions, which exists on some intangible plane between the measurable worlds of oxytocin levels in a cat's bloodstream and the cat's purring? Why not refrain from hypothesizing the cat's happiness? The answer is that emotions are in a real sense where we live, what we care about. Human life cannot be understood without emotions. To leave questions of animal emotion as forever unapproachable and imponderable is arbitrary intellectual helplessness.

Across the Species Barrier

In January of 1989 hikers in the Michigan woods found a black bear with two cubs, recently out of hibernation, curled under a tree. They began taking photographs, and when the bear seemed insufficiently lively for their artistic purposes they shouted and prodded her with sticks. She ran away, leaving behind her twelve-week-old cubs.

Rangers tracked the mother and decided she was not coming

back. Wildlife biologist Lynn Rogers, in Minnesota's Superior National Forest, agreed to try to get the cubs adopted. Carrying Gerry, the female cub, Rogers and a photographer snowshoed into the woods and located Terri, a wild bear with two cubs, who was habituated to human presence. Rogers produced the squalling cub. "I tossed it to her and immediately she wanted it," he recalls. The cub ran from the strange bear, back to the humans. To the photographer's horror, she climbed his leg like a tree. As he stood frozen, Terri walked over, took the cub in her mouth, peeled her off his leg, and carried her back to the den.

The adoption of Gerry's brother by another bear was also successful. Terri was a good mother, and Gerry rambled the north woods, learning to forage—breaking into ant hills, traveling forty miles to a hazelnut stand, grazing on aquatic vegetation—and sleeping under a pine. She grew up to use part of Terri's territory and to have cubs of her own.

During a period when Rogers was at odds with government agencies, officials accused Gerry of attacking humans. She was captured, with one cub. Caged, Gerry moaned constantly. "She just was crying all the time," Rogers says. "Then when we caught the other cubs and put them in the cage with her, she was fine from that moment." Officials planned to ship Gerry to a game farm, where she would be used to breed cubs for sale, and where her toes would be clipped off for safety reasons. Appalled, Rogers managed to arrange for her to go to a small zoo, where she lives in a several-acre enclosure. When her cubs were old enough, they were released in a North Carolina forest.

"That bear was so trustworthy, even with cubs," Rogers mourns. "I could scoop her up in my arms . . . and she'd just relax and look around." As for Terri, she ranged into unprotected forest and was shot by a hunter. In this story the sources of tragedy have nothing to do with bears and everything to do with mistakes that only humans make. The emotional lives of these bears are not inaccessible to us. To deny the terror of the abandoned bear cub; the welcoming affection of the adoptive mother, Terri; Gerry's despair when two of her cubs were missing, is to defy credibility.

Curiosity about the feelings of animals, which science so often tries to train out of its students, may actually be reciprocated by animals. When observing wild lions, Elizabeth Marshall Thomas discovered that the lions were returning human scrutiny. During the day the scientists watched the lions sleep. At night, tracks revealed that four lions came to the fence and peered at the sleeping scientists. As the people examined the scats of the lions, the lions dug up the human latrine and inspected its contents, sometimes adding their own. Wild chimpanzees who have overcome their fear of humans show considerable curiosity about human behavior, though none seems to have gone so far as to make a career out of it.

In the end, when we wonder whether to ascribe an emotion to an animal, the question to ask is not, "Can we prove that another being feels this or any emotions?" but rather, "Is there any reason to suppose that this species of animal does *not* feel this emotion?" If not, then we can ask whether the individual animal feels that particular emotion in this particular instance. If we see an elephant standing with another, dying elephant, the appropriate response is not to say that we have no way to measure sorrow and must therefore never speak of sadness in elephants. Instead we can observe the behavior of the elephant—its calls, body language, and actions—and ask whether it does seem to evidence unhappiness. The animal's personal story is relevant to this inquiry—were these animals strangers? Acquainted? Family? Even if animals do not (as far as we know) tell stories, they surely live them every bit as much as humans do.

Scientific humility suggests that complete understanding of other animals may be impossible. But we will come far closer if we do not begin by insisting that we already know more than we do about what characteristics they do not have. To learn about other animals, they must be taken on their own terms, and these terms include their feelings.

Conclusion:
Sharing the World
with Feeling Creatures

Jeffrey Moussaieff Masson

What are the implications of finding that animals lead emotional lives? Must we change our relationships with them? Have we obligations to them? Is testing products for humans on animals defensible? Is experimentation on animals ethical? Can we confine them for our edification? Kill them to cover, sustain, and adorn ourselves? Should we cease eating animals who have complex social lives, are capable of passionate relations with one another and desperately love their children?

Humans often behave as if something like us were more worthy of respect than something not like us. Racism can partly be described, if not explained, in this way. Men treat other men better than they treat women, based in part on their view that women are not like them. Many of these so-called differences are disguises for whatever a dominant power can impose.

The basic idea seems to be that if something does not feel pain in the way a human being feels pain, it is permissible to hurt it. Even though this is not necessarily true, the illusion of differences is maintained out of fear that seeing similarity will create an obligation to accord respect and perhaps even equality. This appears to be the case especially when it comes to suffering, pain, sorrow, sadness. We do not want to cause these things in others because we know what it feels like to experience them ourselves. No one defends suffering as such. But animal experimentation? The arguments revolve around utility, pitting the greater good against the lesser suffering. Implicit, usually, is the greater importance of those who stand to gain (for example, the scientists employed by cosmetic or pharmaceutical companies to do experiments on rabbits) compared with the lesser importance of those who are sacrificed to their benefit.

An animal experimenter will almost inevitably deny that animals suffer in the same way humans do. Otherwise he would implicitly admit to cruelty. Experimental suffering is not randomly imposed without consent on human beings and defended as ethical on the grounds that it would bestow enormous benefit to others. (At least not any longer.) Animals suffer. Can we, should we, measure their suffering, compare it to our own? If it is like ours, how can it continue? As Rousseau wrote in *Discourse on the Origin of Inequality* in 1755: "It seems that, if I am obliged not to injure any being like myself, it is not so much because he is a reasonable being, as because he is a sensible being." Moreover, why should the suffering have to be like ours to be unjustifiable to inflict? It has been argued that humans experience pain more acutely because we remember and anticipate it; in Rousseau's term we are "reasonable." Yet it is not apparent that animals cannot do both.

But even if they cannot remember or anticipate pain, there is no reason to suppose that they suffer any less than humans do—they are "sensible"—while there is some reason to suppose that some may suffer more. British philosopher Brigid Brophy, for example, points out that "pain is likely to fill the sheep's whole capacity for experience in a way it seldom does in us, whose intellect and imagination can create breaks for us in the immediacy of our sensations." But isn't the fact that they suffer at all enough? Speaking of the connection between suffering and selfless love in animals, Darwin wrote: "In the agony of death a dog has been known to caress his master, and every one has heard of the dog suffering under vivisection, who licked the hand of the operator; this man, unless the operation was fully justified by an increase of our knowledge, or unless he had a heart of stone, must have felt remorse to the last hour of his life." As to animals, he spoke from observation. As to humans, he was optimistic.

It is often said that if slaughterhouses were made of glass, most people would be vegetarians. If the general public knew what went on inside animal experimentation laboratories, they would be abolished. However, the parallel is not exact. Slaughterhouses are invisible because the public does not want to see them. Everyone knows what goes on inside them; they simply do not want to be confronted with it. Most people do *not* know how animals are used in experiments. Slaughterhouses allow visitors. Laboratories where animal experiments are performed are notoriously secretive, off-limits to visitors. Perhaps those who conduct the experiments know they would be stopped if what they did was known even by other scientists. Perhaps they are ashamed. Dr. Robert White, director of the Neurological and Brain Research Laboratory at Cleveland's Metropolitan General Hospital, is a leading figure in brain transplant research. In an influential article entitled "A Defense of Vivisection" he describes his own research: "In 1964, we were successful for the first time in medical history in totally isolating the subhuman primate brain outside of its body and sustaining it in a viable state by connecting it with the vascular system of another monkey or with a mechanical perfusion circuit that incorporated engineering units designed to perform the functions of the heart, lungs and kidneys while simultaneously circulating blood to and from the brain. We were overjoyed since scientists had attempted to construct such a model surgically for the last one hundred years without success. As late as the 1930s Dr. Alexis Carrel, the Nobel laureate, with the collaboration of Colonel Charles Lindbergh, had been able to support the viability of almost all body organs in an isolated state. . . . Parenthetically, it should be mentioned that Dr. Carrel had his problems with the antivivisectionists of his time."

One animal experimentation group ran a paid advertisement in a newspaper, one that they saw as amusing, appealing for donations: "Send a mouse to college." The language disguises the purpose of mice in the university. The experimenters dare not say: "Grow a tumor on a mouse." Nor do they dare to say: "Send a cat or a dog to college," since people do not like to think of their pets as the subjects of experiments. Rats and mice are not generally regarded as pets, but as pests; they have few defenders. Yet the pain a rat or a mouse feels is every bit as real as that of any pet. In laboratories they suffer, as anybody who has heard them moan, cry, whimper, and even scream knows. Scientists dissimulate about this by insisting that they are merely vocalizing. Descartes lives on.

Perhaps these sounds fail to reach scientists because they are not immediately recognizable as a form of communication. In examining the human view of differences between humans and animals it is clear that humans assign primary significance to speech. Our glorious uniqueness, many philosophers have claimed, lies in our ability to speak to one another. It thus came as a shock to learn that a simple African grey parrot not only "parroted" human speech, but spoke, communicated—the words used meant something. When animal psychologist Irene Pepperberg turned to leave her parrot, Alex, in a veterinarian's office for lung surgery, Alex called out, "Come here. I love you. I'm sorry. I want to go back." He thought he had done something bad and was being abandoned as punishment. Imagine what would happen if an animal addressed us on its imminent murder. If, in a slaughterhouse, a pig cried out: "Please don't kill me." If, as a hunter looked into the eyes of a deer, it suddenly broke into speech: "I want to live, please don't shoot, my children need me." Would the hunter pull the trigger? Or if a cat in a laboratory were to cry out: "Please, no more torture," would the scientist be able to continue? Such speech did not stop concentration camp inmates from being murdered during the Holocaust; there, humans, it was said, were lice and rats.

No one assumes the pig wants to die. It would avoid slaughter if it could. It *feels* the desire to live and the pain of its sorrow in being killed just as humans do; the only difference is that it cannot say so in words. The crying of the pigs being slaughtered is horrible. People report that they sound like human screams. The pigs are communicating their terrible fear. Recently a steer on the way to the slaughterhouse was reported to have bolted when it was close enough to hear the anguished cries of the animals. It fled through the town like a prisoner condemned to death. Its sudden lunge for freedom gave everyone pause, even the driver of the death caravan. Was it right to send an animal to slaughter who so desperately wished to live? Perhaps just this one could be saved. Then what about the others? Do they feel the same way? If resistance is to be respected, does lack of resistance confer a right to kill? We *do* know what the cow wants: the cow wants to live. The cow does not wish to sacrifice itself for any reason. That a cow will willingly offer itself as food is a fable.

When humans refuse to inflict pain on other humans, surely it is because they assume they *feel*. It is not because another person can think, nor because they can reason, nor even because they can speak that we respect their physical boundaries, but because they feel. They feel pain, humiliation, sorrow, and other emotions, perhaps even some we do not yet recognize. We do not want to cause suffering. If, as I believe, animals feel pain and sorrow and all the other emotions, these feelings cannot be ignored in our behavior toward them. A bear is not going to compose Beethoven's Ninth Symphony, but then neither is our next door neighbor. We do not for this reason conclude that we have the freedom to experiment upon him, hunt him for sport, or eat him for food.

Modern philosophers seem somewhat more willing than biologists to consider animal emotions, and they have also become engaged in issues of animal rights. Philosophers like Mary Midgley and Brigid Brophy in England, Peter Singer in Australia, and Tom Regan and Bernard Rollin in the United States, all take a strong position that animals are capable of complex emotions. In an influential passage Jeremy Bentham in 1789 connected sentient feelings with rights this way:

> The day *may come*, when the rest of the animal creation may acquire those rights which never could have been withholden from them but by the hand of tyranny. The French have already discovered that the blackness of the skin is no reason why a human being should be abandoned without redress to the caprice of a tormentor. It may come one day to be recog-

nized, that the number of the legs, the villosity of the skin, or the termination of the *os sacrum*, are reasons equally insufficient for abandoning a sensitive being to the same fate. What else is it that should trace the insuperable line? Is it the faculty of reason, or, perhaps, the faculty of discourse? But a full-grown horse or dog is beyond comparison a more rational, as well as a more conversable animal, than an infant of a day, or a week, or even a month, old. But suppose the case were otherwise, what would it avail? The question is not, Can they *reason?* nor, Can they *talk?* but, Can they *suffer?*

Peter Singer, in *Animal Liberation,* explicitly based on Bentham's nineteenth-century utilitarianism, argues that creatures that can feel pain deserve to be shielded from that pain, especially from scientific experimentation and hurtful farming methods. The argument is that sentience—the capacity to have conscious experiences —demands equal consideration to the interests of all creatures. However, although this provides one moral ground, this position does not explicitly accord animals rights.

Tom Regan in *The Case for Animal Rights* goes further, arguing explicitly for protecting the rights of animals who are "capable of being the subject of a life." Every animal used in every experiment in every laboratory has its own life story. It has felt strong emotions, loved and hated and been devoted to others of its own kind. It is a subject, and is therefore violated by being treated as an object. Have we the right to tear this being away from its fellows and all that gives its life meaning and put it in a sterile, hostile, aseptic environment to be tortured, maimed, and ultimately destroyed in the name of anything, far less of service to our species? Or lacking the right, do we only have the power?

What is learned from these experiments is not always of benefit to humans. It was recently reported in a German psychiatric journal that a researcher gave Largactil, a neuroleptic tranquilizer, to a spider, and succeeded either in diminishing the size and complexity of its web, or in stopping the spider from spinning a web at all. This article was held up as evidence of the great value of animal research in psychology. It meant, said the researcher, that antipsychotic drugs can be given to schizophrenics to stop them from spinning webs, that is, from creating fantasies in their heads. But why should spiders, or humans for that matter, not spin webs if so inclined? Who bestowed upon us the right to intervene and interfere and ultimately destroy the delicate product of a creature's innermost being? Whether such practices ultimately enhance humanity is also questionable. The microbiologist Catherine Roberts condemns Harry Harlow's "odious" experiments on rhesus monkeys (discussed in Chapter 5) pointing out that they "degrade the humanness of those who designed and perpetrated them." Dr. Roberts also had a comment to make about Dr. White's experiments on brain transplantation. She said: "The details of his experiments are so horrifying that they seem to reach the limits of scientific depravity."

It may be hard to imagine the sensual universe of another species, but it is not impossible. Our dog's intense sniffing suggests she is picking up and responding to something beyond our ken. Her ability to take in information hidden from us is impressive; the resulting sudden shifts of mood are honored. We know we are in the presence of something different from us but worthy of our respect. One of the most common emotions humans feel in the presence of another species is awe. The ability of a hawk to soar, of a seal to race through the waves is marvelous, humbling.

It is clear that animals form lasting friendships, are frightened of being hunted, have a horror of dismemberment, wish they were back in the safety of their den, despair for their mates, look out for and protect their children whom they love. As Tom Regan would say, they are the subject of lives, as we are. Though animals do not write autobiographies, as we understand them, their biographies can be written. They are individuals and members of groups, with elaborate histories that take place in a concrete world, and involve a large number of complex emotional states. They *feel* throughout their lives, just as we do.

Jane Goodall points out that "chimpanzees differ genetically from *Homo sapiens* by only about 1 per cent, and that while they lack speech, they nevertheless behave similarly to humans, can feel pain, share our emotions and have sophisticated intellectual abilities." She pleads that we stop enslaving, imprisoning, incarcerating, and torturing them, and instead protect them from exploitation.

"If I learned anything from my time among the elephants," writes the scientist Douglas Chadwick:

> it is the extent to which we are kin. The warmth of their families makes me feel warm. Their capacity for delight gives me joy. Their ability to learn and understand things is a continuing revelation for me. If a person can't see these qualities when looking at elephants, it can only be because he or she doesn't want to.

Humans have long recognized that animals have the potential to connect emotionally with humans. One of the oldest and most popular Indian tales is about the life-and-death bond between a Brahmin and a mongoose. Here it is as found in the great *Ocean of Story*, written about A.D. 1070, a Kashmiri collection: "A Brahmin by the name of Devasharman lived in a certain village. He had a wife of equally high birth, named Yajnadatta. She became pregnant, and in time gave birth to a son. The Brahmin, though poor, felt he had obtained a great gem. After she had given birth to the child, the Brahmin's wife went to the river to bathe. Devasharman remained in the house, taking care of his infant son. Meanwhile a maid came from the women's apartments of the palace to summon the Brahmin, who lived on presents received by performing religious ceremonies. . . . To guard the child, he left a mongoose, which he had raised in his house since it was born. As soon as the Brahmin left, a snake suddenly slithered toward the child. The mongoose, seeing it, killed it out of love for his master.

In the distance, the mongoose saw Devasharman returning. Happy to see him, he ran towards him, stained with the blood of the snake. But when Devasharman saw the blood, he thought: "Surely he has killed my little boy," and in his delusion he killed the mongoose with a stone. When he went into the house he saw the snake killed by the mongoose and his boy alive and safe. He felt a deep inner sorrow. When his wife returned and learned what had happened, she reproached him, saying: "Why did you not think before killing this mongoose which had been your friend?"

This is what Jan Harold Brunvand (*The Choking Doberman*) calls an "urban legend." He writes about the age-old folk fable of helpful animals: "A classic European manifestation of this legend is the Welsh 'Llewellyn and Gellert,' in which the faithful hunting hound Gellert is found bloodied and gasping in the hall of Prince Llewellyn's home. The dog is presumed to have killed the baby it was left to guard, whose overturned crib is seen through the open doorway. The dog is slain, but the baby is found unharmed; and the hidden intruder that Gellert had defended the infant from—a huge wolf—is found inside the house dead from the dog's defensive efforts." He notes: "Although revered by many in Wales as an ancient national legend—or even as history—the story of Llewellyn and Gellert is, as Welsh historian Prys Morgan phrased it, 'of course all moonshine, or more exactly, a clever adaptation of a well-known international folktale.' "

We cannot know whether the events really happened. The story is not so highly improbable. Mongooses are often kept as pets in India, and mongooses do in fact prey upon snakes, including cobras and other highly venomous species. But whether or not based on fact, such accounts exert a powerful hold on the imaginations of many different cultures: versions are found in Mongolian, Arabic, Syriac, German, an English ballad of William R. Spencer's, and others. They clearly speak to a sense of animal loyalty and clarity, of human arrogance and guilt, an awareness of the precariousness of human judgments. Can we be trusted to honor the deep bond that a dog or mongoose can form with us? The "legend," if that is what it is, speaks better for animals than humans.

Perhaps the most famous account testifying at least to the hope, and possibly the fact, of a bond of gratitude, friendship, and compassion between a person and an animal is the ancient account of Androcles and the lion. An early recorded version in Latin appears in the *Attic Nights of Aulus Gellius* in the second century. The account is prefaced with a claim to authenticity: "The account of Apion, a learned man who was surnamed Plistonices, of the mutual recognition, due to old acquaintance, that he had seen at Rome between a man and a lion. . . . This incident, which he describes in the fifth book of his *Wonders of Egypt*, he declares that he neither heard nor read, but saw himself with his own eyes in the city of Rome." Gellius then quotes Apion:

> In the Great Circus a battle with wild beasts on a grand scale was being exhibited to the people. Of that spectacle, since I chanced to be in Rome, I was an eyewitness. There were many savage wild beasts, brutes remarkable for their huge size, and all of uncommon appearance or unusual ferocity. But beyond all others did the vast size of the lions excite wonder, and one of these in particular surpassed all the rest because of the huge size of his body. . . . There was brought in . . . the slave of an ex-consul; the slave's name was Androcles. When that lion saw him from a distance he stopped short as if in amazement, and then approached the man slowly and quietly, as if he recognized him. Then, wagging his tail in a mild and caressing way, after the manner and fashion of fawning dogs, he came close to the man, who was now half dead from fright, and gently licked his feet and hands. . . . Then you might have seen man and lion exchange joyful greetings, as if they had recognized each other.

The emperor Caligula wanted to know why the lion had spared the man. Androcles related how he had run away from his master into the lonely desert and hidden in a remote cave. A lion came into the cave with a bleeding paw, groaning and moaning in pain. The lion, Androcles is reported to have said, "approached me mildly and gently, and lifting up his foot, was evidently showing it to me and holding it out as if to ask for help." Androcles took out a huge splinter and cared for the foot. "Relieved by that attention and treatment of mine, the lion, putting his paw in my hand, lay down and went to sleep." For three years they shared the cave, the lion hunting for both. Then Androcles was recaptured, returned to

Rome, and condemned to death in the arena. Upon hearing this story, Caligula, after a vote by the people, freed lion and man. They walked the streets together "and everyone who met them anywhere exclaimed: 'This is the lion that was a man's friend, this is the man who was physician to a lion.' "

Is this fiction, testimony to an ancient longing in the human heart to love and be loved by another animal as one longs to love and be loved by another person? It is not so far removed from Joy Adamson's account of the lion Elsa, whom she raised and then released; for years afterward Elsa returned from the wild to visit with her children and her mate.

Reciprocity on the level of Androcles and the lion, this dream of equality, may be closed to us for now. But whether or not it can be realized, we do owe animals something. Freedom from exploitation and abuse by humankind should be the inalienable right of every living being. Animals are not there for us to drill holes into, clamp down, dissect, pull apart, render helpless, and subject to agonizing experiments. John Lilly, one of the first to work scientifically with dolphins, was recently quoted as saying that he no longer works with dolphins because he "didn't want to run a concentration camp for highly developed beings." Animals are, like us, endangered species on an endangered planet, and we are the ones who are endangering them, it, and ourselves. They are innocent sufferers in a hell of our making. We owe them, at the very least, to refrain from harming them further. If no more, we could leave them be.

When animals are no longer colonized and appropriated by us, we can reach out to our evolutionary cousins. Perhaps then the ancient hope for a deeper emotional connection across the species barrier, for closeness and participation in a realm of feelings now beyond our imagination, will be realized.

Notes

Prologue: Searching the Heart of the Other

xvii Charles Darwin, *The Expression of the Emotions in Man and Animals* (1872; reprint, Chicago and London: University of Chicago Press, 1965).

xvii " 'Who can say . . .' " Charles Darwin, *The Descent of Man; and Selection in Relation to Sex* (1871; reprint, Princeton: Princeton University Press, 1981), pp. 62, 76. Also see the discussion of animal emotions in J. Howard Moore, *The Universal Kinship* (1906; reprint, Sussex, England: Centaur Press, 1992).

xvii Donald Griffin, *The Question of Animal Awareness: Evolutionary Continuity of Mental Experience* (New York: Rockefeller University Press, 1976). Griffin is the discoverer of bat sonar. In the bibliography are listed those of his books and articles that affected the thinking in this book.

xviii " 'When, in the early . . .' " Paola Cavalieri and Peter Singer, eds., *The Great Ape Project: Equality Beyond Humanity* (London: Fourth Estate, 1993), p. 12.

xviii " 'A lion is not . . .' " George Adamson, *My Pride and Joy* (New York: Simon & Schuster, 1987), p. 19.

xx "Comparative psychology to this day . . ." Thus the *Journal of Comparative Psychology* announces in each issue that it publishes "research on the behavior and cognitive abilities of different species (including humans) as they relate to evolution, ecology, adaptation, and development. Manuscripts that focus primarily on issues of proximate causation where choice of specific species is not an important component of the research fall outside the scope of this journal."

xx ". . . unworthy of scientific attention." In a much discussed article in *Der Spiegel* (Nr. 47, 1980, pp. 251–62) entitled "Tiere sind Gefühlsmenschen" [Animals Are Feeling Creatures], Konrad Lorenz speaks of "crimes against animals" and says that anybody who intimately knows any individual higher mammal such as a dog or an ape and does *not* believe that this creature has feelings similar to his own is crazy. ("Ein Mensch, Der Ein höheres Säugetier, etwa einen Hund oder einen Affen, wirklich genau kennt und *nicht* davon überzeugt wird, dass dieses Wesen ähnliches erlebt wie er selbst, ist psychisch abnorm . . .")

xxi E. Sue Savage-Rumbaugh, *Ape Language: From Conditioned Response to Symbol* (New York: Columbia University Press, 1986), p. 25.

Chapter 1: In Defense of Emotions

2 G. G. Rushby, "The Elephant in Tanganyika," in Ward, Rowland, *The Elephant in East Central Africa: a Monograph* (London and Nairobi: Rowland Ward Ltd., 1953). Cited in Richard Carrington, *Elephants: A Short Account of Their Natural History, Evolution and Influence on Mankind* (London: Chatto & Windus, 1958), p. 83.

2 " 'arguably the most important . . .' " Savage-Rumbaugh, *Ape Language: From Conditioned Response to Symbol*, p. 266.

3 Jane Goodall, interview by Susan McCarthy, May 7, 1994.

4 Mary Midgley, "The Mixed Community," in Eugene C. Hargrove, ed., *The Animal Rights/Environmental Ethics Debate* (Albany: State University of New York Press, 1992), p. 214.

4 Gunther Gebel-Williams with Toni Reinhold, *Untamed: The Autobiography of the Circus's Greatest Animal Trainer* (New York: William Morrow & Co., 1991), p. 28.

4 "trainers were startled . . ." Personal communication, August 23, 1994.

5 " 'A loving dog-owner . . .' " In Schaller's foreword to Shirley Strum, *Almost Human: A Journey into the World of Baboons* (New York: Random House, 1987), p. xii.

5 " 'Intuitively I seemed . . .' " George and Lory Frame, *Swift & Enduring: Cheetahs and Wild Dogs of the Serengeti* (New York: E.P. Dutton, 1981), p. 156.

6 Anne Rasa, *Mongoose Watch: A Family Observed* (Garden City, NY: Anchor Press/Doubleday & Co., 1986).

7 "Female baboons kept together . . ." Thelma Rowell, *The Social Behaviour of Monkeys* (Harmondsworth, Middlesex, England: Penguin, 1972), p. 79.

7 Hope Ryden, *God's Dog* (New York: Coward, McCann & Geoghegan, 1975), pp. 87, 92–101.

8 "The female Tasmanian . . ." J. Maynard Smith and M. G. Ridpath, "Wife Sharing in the Tasmanian Native Hen, *Tribonyx mortierii:* A Case of Kin Selection?" *The American Naturalist* 106 (July–August 1972), pp. 447–52.

8 " 'There are willing workers . . .' " Robert Cochrane, "Working Elephants

at Rangoon," quoted in *The Animal Story Book*, Vol. IX, The Young Folks Library (Boston: Hall & Locke Co., 1901).

8 "Theodore Roosevelt . . ." Quoted in Paul Schullery, *The Bear Hunter's Century* (New York: Dodd, Mead & Co., 1988), p. 142.

9 David McFarland, ed., *The Oxford Companion to Animal Behavior* (Oxford and New York: Oxford University Press, 1987), p. 151.

9 " 'It is surely . . .' " Quoted by Sydney E. Pulver in an excellent overview of the topic: "Can Affects Be Unconscious?" *International Journal of Psycho-Analysis*, 52 (1971), p. 350.

10 "*alexithymia* . . ." Robert Jean Campbell, *Psychiatric Dictionary*, 5th ed. (New York and Oxford: Oxford University Press, 1981), p. 24.

10 "Psychological theorists speak . . ." Carroll Izard and S. Buechler, "Aspects of Consciousness and Personality in Terms of Differential Emotions Theory," in *Emotion: Theory, Research, and Experience, Vol. I: Theories of Emotion*, Robert Plutchik and Henry Kellerman, eds. (New York: Academic Press, 1980), pp. 165–87.

10 "One psychologist compiled . . ." Joseph de Rivera, *A Structural Theory of the Emotions* (New York: International Universities Press, 1977), pp. 156–64.

10 "William James defined . . ." June Callwood, *Emotions: What They Are and How They Affect Us, from the Basic Hates and Fears of Childhood to More Sophisticated Feelings That Later Govern Our Adult Lives: How We Can Deal with the Way We Feel* (Garden City, NY: Doubleday & Co., 1986), p. 33.

10 "Behaviorist J. B. Watson . . ." Robert Thomson, "The Concept of Fear," in *Fear in Animals and Man*, W. Sluckin, ed., 1–23 (New York and London: Van Nostrand Reinhold Co., 1979), pp. 20–21.

10 "Modern theorists . . ." Michael Lewis, *Shame: The Exposed Self* (New York: The Free Press/Macmillan, 1992), pp. 13–14.

11 Anna Wierzbicka, "Human Emotions: Universal or Culture-Specific?" *American Anthropologist* 88 (1986), pp. 584–94.

11 Lévy-Bruhl, *Les fonctions mentales dans les sociétés inférieures* (Paris: Félix Alcan, 1910). It was published in the Bibliothèque de Philosophie Contemporaine, under the direction of Emile Durkheim. Lévy-Bruhl contrasts the primitive mentality with that of the "individu blanc, adulte et civilisé" [the white, adult and civilized individual] (p. 2). One proof (p. 31): Cherokee Indians believe that "fish live in a civil society like men, and have villages and roads in the water." These same "primitives" believe in expiatory rites before killing animals (p. 32). Moreover, "they" cannot generalize, and "every species of monkey and palm tree has its own name" (p. 192) and we "must not be led into believing that these delicate distinctions in the same species of plant or animals demonstrates an interest in objective reality" (p. 198). The text was much read and cited for many years.

13 Gordon M. Burghardt, "Animal Awareness: Current Perceptions and Historical Perspective," *American Psychologist* 40 (August 1985), pp. 905–19.

14 Frans de Waal, *Peacemaking Among Primates* (Cambridge, MA and London: Harvard University Press, 1989), p. 25.

15 "Grossly oversimplified . . ." For an account of the problems with correlating testosterone levels and aggressiveness, for example, see Alfie Kohn, *The*

Brighter Side of Human Nature: Altruism and Empathy in Everyday Life (New York: Basic Books, 1990), pp. 27–28.

16 "The part of the brain . . ." See Gordon G. Gallup, Jr. and Susan D. Suarez, "Overcoming Our Resistance to Animal Research: Man in Comparative Perspective," in *Comparing Behavior: Studying Man Studying Animals*, D. W. Rajecki, ed. (Hillsdale, NJ: Lawrence Erlbaum Associates, 1983), p. 10. They note: "The basic biological principles governing the metabolic, endocrinological, neurological, and biochemical activities in man are basically the same in many other organisms. Behavior, therefore, has become the last stronghold for the Platonic paradigm. . . . If we accept the proposition that, in the last analysis, behavior is nothing more than an expression of physiological processes, then to admit the biological but deny the psychological similarities between ourselves and other species seems logically inconsistent and indefensible."

18 Descartes is quoted in Tom Regan and Peter Singer, eds., *Animal Rights and Human Obligations* (New Jersey: Prentice-Hall, 1979), pp. 61–64. The original passage is from *Discours de la méthode*, 5 (A. Bridoux, ed., *Oeuvres et lettres de Descartes*, pp. 165–66. Dijon, France: Gallimard, 1953).

18 "An unknown contemporary . . ." Quoted in Tom Regan, *The Case for Animal Rights* (Berkeley, CA: University of California Press, 1983), p. 5.

18 " 'Answer me . . .' " François-Marie Arouet de Voltaire, *Dictionnaire philosophique*, Julien Benda and Raymond Naves, eds. (Paris: Garnier Frères, 1961), pp. 50–51. Translation by Jeffrey Masson.

18 "Elsewhere . . ." François-Marie Arouet de Voltaire, "The Beasts," Article 6 in *Le philosophe ignorant*, Les Oeuvres Complètes de Voltaire, Vol. Mélanges, Jacques van den Heuvel, ed. (Paris: Gallimard), p. 863. Translation by Jeffrey Masson.

18 "As early as 1738 . . ." The French text is quoted in Hester Hastings, *Man and Beast in French Thought of the Eighteenth Century*, Vol. 27 (Baltimore: The Johns Hopkins Press, 1936), p. 183. Translation by Jeffrey Masson. See, too, Thomas H. Huxley, "On the hypothesis that animals are automata, and its history," in *Method and Results: Essays* (1893; reprint, London: Macmillan, 1901), pp. 199–250. He writes, "I confess that, in view of the struggle for existence which goes on in the animal world, and of the frightful quantity of pain with which it must be accompanied, I should be glad if the probabilities were in favour of Descartes' hypothesis; but, on the other hand, considering the terrible practical consequences to domestic animals which might ensue from any error on our part, it is as well to err on the right side, if we err at all, and deal with them as weaker brethren, who are bound, like the rest of us, to pay their toll for living, and suffer what is needful for the general good. As Hartley finely says, 'We seem to be in the place of God to them.' " (*Ibid.*, p. 237) For a complete history of the Descartes debate, see Leonora Cohen Rosenfield: *From Beast-Machine to Man-Machine: Animal Soul in French Letters from Descartes to La Mettrie* (1940; new edition, New York: Octagon Books, 1968); the introduction in François Dagognet, "L'Animal selon Condillac" in *Traité des animaux* (Paris: Librairie Philosophique J. Vrin, 1987); and George Boas, "The Happy Beast" in *French Thought of the Seventeenth*

Century: Contributions to the History of Primitivism (Baltimore: The Johns Hopkins Press, 1933).

19 Irene Pepperberg, interview by Susan McCarthy, February 22, 1993.

21 Elizabeth Marshall Thomas, "Reflections: The Old Way," *The New Yorker* (October 15, 1990), p. 91.

21 De Waal, *Peacemaking Among Primates*, p. 220.

21 David Macdonald, *Running with the Fox* (London and Sydney: Unwin Hyman, 1987), p. 164.

21 Konrad Lorenz, *The Year of the Greylag Goose* (New York and London: Harcourt Brace Jovanovich, 1978), p. 56.

22 "Not only is it . . ." Cf. Mary Midgley, *Beast and Man: The Roots of Human Nature* (Ithaca, NY: Cornell University Press, 1978), p. 345.

Chapter 2: Unfeeling Brutes

25 *Oeuvres choisies de Buffon, Vol. 2:* "L'Histoire naturelle des animaux" (Paris: Librairie de Firmin Didot Frères, 1861), pp. 484–88, 493–96, 509, 525.

26 N. K. Humphrey, "The Social Function of Intellect," in *Growing Points in Ethology*, P. P. G. Bateson and R. A. Hinde, eds., pp. 303–17 (Cambridge, England: Cambridge University Press, 1976).

26 Donald Symons, *The Evolution of Human Sexuality* (New York: Oxford University Press, 1979), pp. 78–79.

26 "When the question . . ." D. Goldfoot et al., "Behavioral and Physiological Evidence of Sexual Climax in the Female Stump-tailed Macaque," *Science* 208 (1980), pp. 1477–79. Cited in de Waal, *Peacemaking Among Primates*, pp. 151–53.

26 De Waal, *Peacemaking Among Primates*, pp. 151–53, 198–206.

27 " 'Civilization, or perhaps . . .' " This splendid example of benightedness is also quoted by Mary Midgley in her article "The Mixed Community," in Hargrove, ed., *The Animal Rights/Environmental Ethics Debate*, p. 223. The actual article is a very long and learned one written by Northcote W. Thomas, in Vol. 1 of the *Encyclopedia of Religion and Ethics*, edited by James Hastings (Edinburgh: T. & T. Clark, 1908), pp. 483–535. The article actually begins (p. 483) by citing the "great gulf that exists between man . . . and the elephant and the anthropoid ape."

28 Matt Cartmill, *A View to a Death in the Morning: Hunting and Nature Through History* (Cambridge, MA: Harvard University Press, 1993), p. 222.

28 Stephen Jay Gould, *The Mismeasure of Man* (New York: W.W. Norton and Co., 1981).

29 " 'The fact that . . .' " Volker Arzt and Immanuel Birmelin, *Haben Tieren ein Bewusstsein?: Wenn Affen lügen, wenn Katzen denken und Elefanten traurig sind* (Munich: C. Bertelsmann, 1993), p. 154. Translation by Jeffrey Masson.

29 "When the subject . . ." "Another assessment of pain in fish comes from a team of researchers under the direction of Professor John Verheijen at the University of Utrecht in the Netherlands [in 1988]. They concluded that fish do feel pain and experience fear." R. Barbara Orleans, *In the Name of Science:*

Issues in Responsible Animal Experimentation (New York: Oxford University Press, 1992), p. 148.

29 "The history . . ." E. S. Turner recently commented about his 1964 book *All Heaven in a Rage*, one of the first books to challenge attitudes toward animals: "In my original introduction I commented that in our attitude to animals we are hopelessly, perversely, inconsistent. Reviewing this book in the *Observer*, Philip Toynbee followed up the point, remarking that the rage of English foxhunters knew no bounds when they learned that the Russians had shot a dog into space. He thought that a certain pattern could be traced in these bewildering inconsistencies. 'We abominate the cruelties which we are not tempted to perform, and we abominate them all the more when they are practiced by people who do not belong to our own group.' He could have added 'or when they are practiced by people of another nation.' " E. S. Turner, *All Heaven in a Rage* (Sussex, England: Centaur Press, 1992), pp. 323–24.

29 "Similarly, until the 1980s . . ." This incredible practice is well attested to in medical sources. See K. J. S. Anand and P. J. McGrath, eds., *Pain in Neonates* (Amsterdam: Elsevier, 1993); Neil Schechter, Charles B. Berde, and Myron Yaster, eds., *Pain in Infants, Children, and Adolescents* (Baltimore: Williams and Wilkins, 1993); "Medicine and the Media" (editorial), *British Medical Journal* 295 (September 12, 1987), pp. 659–60; Ian S. Gauntlett; T. H. H. G. Koh; and William A. Silverman, "Analgesia and anaesthesia in newborn babies and infants" (Letters), *Lancet*, May 9, 1987; Nancy Hall, "The Painful Truth" *Parenting* (June/July 1992).

30 "Studies showing . . ." R. N. Emde and K. L. Koenig, "Neonatal Smiling and Rapid Eye-movement States," *Journal of the American Academy of Child Psychiatry* 8 (1969), pp. 57–67. Cited in Carroll Izard, *Human Emotions* (New York and London: Plenum Press, 1977).

32 Article by Frank B. Jevons. Edited by James Hastings. Vol. 1, p. 574.

32 "The philosopher Ludwig Feuerbach . . ." See the article by R. J. Zwi Werblowsky in *The Encyclopedia of Religion* (ed. by Mircea Eliade), Vol. 1, pp. 316–20 (New York: Macmillan, 1987). Old German dictionaries (e.g. *Meyers grosses Konversations-Lexikon* of 1903) speak of anthropopathy, specifically ascribing human emotions to objects and animals (!) that cannot experience them. J. J. Rousseau in *Emile* says: *"Nous sommes pour la plupart de vrais anthropomorphites,"* which may well be, according to the eleventh edition of the Encyclopaedia Britannica, where the term originates.

32 McFarland, ed., *Oxford Companion to Animal Behavior*, p. 17.

32 " 'The scientific study . . .' " John S. Kennedy, *The New Anthropomorphism* (Cambridge, England: Cambridge University Press, 1992), pp. 3–5.

33 " 'anthropomorphism will be . . .' " *Ibid.*, p. 167.

33 John Andrew Fisher, "Disambiguating Anthropomorphism: An Interdisciplinary Review," in *Perspectives in Ethology* 9 (1991), p. 49.

33 "This is one reason . . ." "Male/female differences in attitudes and knowledge of animals were substantial and implied the need for better understanding and appreciation of female attitudes toward and interests in wildlife. Particularly provocative were variations in basic feelings and ethical concern

for animals. The most outstanding result was the much greater humanistic concern for animals among females." Stephen R. Kellert and Joyce K. Berry, *Phase III: Knowledge, Affection and Basic Attitudes Toward Animals in American Society* (U.S. Fish and Wildlife Service, 1980), p. 59. Phase Three gives the results of a U.S. Fish and Wildlife Service funded study of American attitudes, knowledge and behaviors toward wildlife and natural habitats.

34 Frans de Waal, *Chimpanzee Politics: Power and Sex Among Apes* (New York: Harper & Row, 1982), pp. 41–42.

34 Joy Adamson, intro. by J. Huxley, *Living Free* (U.K.: Collins & Harvill Press, 1961), p. xi.

35 Irene Pepperberg, interview by Susan McCarthy, February 22, 1993.

35 "What is wrong with . . ." This theme is also expressed in Theodore Xenophon Barber, *The Human Nature of Birds: A Scientific Discovery with Startling Implications* (New York: St. Martin's Press, 1993).

36 Sy Montgomery, *Walking with the Great Apes* (Boston: Houghton Mifflin, 1991), p. 143.

36 Cynthia Moss, *Elephant Memories: Thirteen Years in the Life of an Elephant Family* (New York: William Morrow and Co., 1988), p. 37.

36 M. Bekoff and D. Jamieson, "Ethics and the Study of Carnivores," in *Carnivore Behavior, Ecology, and Evolution*, 2nd ed. (Ithaca, NY: Cornell University Press, 1995).

36 "Yet as recently as 1987 . . ." Thomas, "Reflections: The Old Way," p. 99.

36 "Bottle-nosed dolphins . . ." Peter Tyack, "Whistle Repertoires of Two Bottle-nosed Dolphins, *Tursiops truncatus:* Mimicry of Signature Whistles?" *Behavioral Ecology and Sociobiology* 18 (1989), pp. 251–57.

36 " 'frequently uttered sounds . . .' " Eberhard Gwinner and Johannes Kneutgen, "Über die biologische Bedeutung der 'zweckdienlichen' Anwendung erlernter Laute bei Vögeln," *Zeitschrift für Tierpsychologie* 19 (1962), pp. 692–96.

37 Mike Tomkies, *Last Wild Years* (London: Jonathan Cape, 1992), p. 172.

37 Mary Midgley, "The Concept of Beastliness: Philosophy, Ethics and Animal Behavior," *Philosophy* 48 (1973), pp. 111–35.

37 Kennedy, *New Anthropomorphism*, p. 87.

38 "How is knowing . . ." "If consciousness has evolved as a biological adaptation for doing introspective psychology, then the presence or absence of consciousness in animals of different species will depend on whether or not they need to be able to understand the behavior of other animals in a social group. Wolves and chimpanzees and elephants, which all go in for complex social interactions, are probably all conscious; frogs and snails and codfish are probably not. . . . The advantage to an animal of being conscious lies in the purely private use it makes of conscious experience as a means of developing an ideology which helps it to model another animal's behavior. It need make no difference at all whether the other animal is actually experiencing the feelings with which it is being credited; all that matters is that its behavior should be understandable on the assumption that such feelings provide the reasons for its actions." N. K. Humphrey: "Nature's Psychologists," in *Consciousness and the Physical World*, B. D. Josephson and V. S.

Ramachandran, eds., 57–80 (Oxford, England: Pergamon Press, 1980), pp. 68–69.

38 "N. K. Humphrey writes . . ." In B. D. Josephson and V. S. Ramachandran, eds., *Consciousness and the Physical World* (Oxford, England: Pergamon Press, 1980), pp. 57–80.

38 Midgley, *Beast and Man: The Roots of Human Nature*, pp. 41, 344–57. Also see Mary Midgley, *Animals and Why They Matter* (Athens, GA: University of Georgia Press, 1983).

39 J. Ortega y Gasset, *Meditations on Hunting*. H. B. Wescott, trans. (New York: Scribner's, 1972), pp. 136–38. Italics in the original. Cited from Matt Cartmill, *A View to a Death in the Morning*, p. 240.

41 "The idea seems to be . . ." One way to avoid such errors has been to pretend that animals are neuter, which has only succeeded in reducing them to things rather than beings. Writing about creative behavior in dolphins for a scientific journal, researcher Karen Pryor was told to call the rough-toothed porpoise Hou "it" rather than "she," on the grounds that "she" should be reserved for referring to humans. Not that being called "she" has humanized women with any security. Refusal to discuss observable facts (Hou was unquestionably female) is hardly scientific. Making the same mistakes about gender in animals that we do in people is no solution to the anthropomorphism taboo. See Karen Pryor, *Lads Before the Wind: Adventures in Porpoise Training* (New York: Harper & Row, 1975), p. 240; Karen Pryor, Richard Haag, and Joseph O'Reilly, "The Creative Porpoise: Training for Novel Behavior," *Journal of the Experimental Analysis of Behavior*, 12 (1969) pp. 653–61.

41 Mike Tomkies, *On Wing and Wild Water* (London: Jonathan Cape, 1987), pp. 136–37.

42 J. E. R. Staddon, "Animal Psychology: The Tyranny of Anthropocentrism," in *Whither Ethology? Perspectives in Ethology*, P. P. G. Bateson and Peter H. Klopfer, eds. (New York: Plenum Press, 1989), p. 123.

42 "Deception has been observed . . ." Robert W. Mitchell and Nicholas S. Thompson, eds., *Deception: Perspectives on Human and Nonhuman Deceit.* (Albany: State University of New York Press, 1986).

43 " 'With a loud grunt . . .' " Jane Goodall, *In the Shadow of Man* (London: Collins, 1971), p. 202. Italics in the original.

43 "One respected random . . ." Diana E. H. Russell, *The Politics of Rape: The Victim's Perspective* (New York: Stein & Day, 1977); Diana E. H. Russell, *Rape in Marriage* (New York: Macmillan, 1982); and Diana E. H. Russell and Nancy Howell, "The Prevalence of Rape in the United States Revisited," *Signs: Journal of Women in Culture and Society* 8 (Summer 1983), pp. 668–95.

43 "Child abuse may . . ." Diana E. H. Russell, "The Incidence and Prevalence of Intrafamilial and Extrafamilial Sexual Abuse of Female Children," *Child Abuse and Neglect: The International Journal* 7 (1983): pp. 133–46; and Diana E. H. Russell, *The Secret Trauma: Incestuous Abuse of Women and Girls* (New York: Basic Books, 1986).

44 Elizabeth Marshall Thomas, *The Hidden Life of Dogs* (Boston and New York: Houghton Mifflin Co., 1993).

Chapter 3: Fear, Hope, and the Terrors of Dreams

45 "And yet from a Kenyan . . ." Douglas H. Chadwick, *The Fate of the Elephant* (San Francisco: Sierra Club Books, 1992), pp. 129, 327.
45 "Wildlife biologist Lynn Rogers . . ." Adele Conover, "He's Just One of the Bears," *National Wildlife* 30 (June–July 1992), pp. 30–36.
46 "Rogers learned . . ." Lynn Rogers, interview by Susan McCarthy, July 15, 1993.
47 "Yet fear has also been . . ." Andrew Mayes, "The Physiology of Fear and Anxiety," in *Fear in Animals and Man*, W. Sluckin, ed., 24–55 (New York and London: Van Nostrand Reinhold Co., 1979), pp. 32–33.
47 McFarland, ed., *Oxford Companion to Animal Behavior*, p. 180.
47 "The biological traces . . ." Melvin Konner, *The Tangled Wing: Biological Constraints on the Human Spirit* (New York: Holt, Rinehart and Winston, 1982), p. 215.
48 "The theory is that . . ." Marcia Barinaga, "How Scary Things Get That Way," *Science* 258 (November 6, 1992), pp. 887–88.
48 "The climber, more often . . ." Thomson, "The Concept of Fear," p. 3.
49 F. Fraser Darling, *A Herd of Red Deer: A Study in Animal Behavior* (London: Oxford University Press, 1937), pp. 70–71.
49 "Somehow it is . . ." Pryor, *Lads Before the Wind*, p. 178.
49 ". . . or that a frightened . . ." *Gorilla: Journal of the Gorilla Foundation* 15, #2 (June 1992), p. 5.
49 " 'We are—not metaphorically . . .' " Konner, *Tangled Wing*, p. 235.
49 Douglas H. Chadwick, *A Beast the Color of Winter: The Mountain Goat Observed* (San Francisco: Sierra Club Books, 1983), pp. 57–58.
49 Wolfgang de Grahl, *The Grey Parrot*, trans. by William Charlton (Neptune City, NY: T.F.H. Publications, 1987), pp. 44–45.
51 "In the Rockies . . ." Chadwick, *Beast the Color of Winter*, p. 89.
51 "Wild birds at a . . ." P. A. Russell, "Fear-Evoking Stimuli," in *Fear in Animals and Man*, W. Sluckin, ed., 86–124 (New York and London: Van Nostrand Reinhold Co., 1979), pp. 97–98.
52 "Wingnut, a particularly . . ." Thomas Bledsoe, *Brown Bear Summer: My Life Among Alaska's Grizzlies* (New York: Dutton, 1987), p. 129.
52 "Separated from his . . ." Pryor, *Lads Before the Wind*, p. 178.
52 Jack Adams, *Wild Elephants in Captivity* (Dominguez Hills, CA: Center for the Study of Elephants, 1981), p. 146.
52 "One aviculturalist . . ." De Grahl, *Grey Parrot*, pp. 210–12.
52 Arjan Singh, *Tiger! Tiger!* (London: Jonathan Cape, 1984), pp. 75, 90.
53 "Cody, an orangutan . . ." Keith Laidler, *The Talking Ape* (New York: Stein and Day, 1980). Laidler was shocked by Cody's terror of his own species and arranged for Cody to meet and eventually be caged with another young orangutan. The two apes became friendly and would walk about hand in hand.
53 Jim Crumley, *Waters of the Wild Swan* (London: Jonathan Cape, 1992), pp. 85–86.
53 Thomas, *Hidden Life of Dogs*, p. 71.

54 "Mountain goats . . ." Chadwick, *Beast the Color of Winter*, p. 115.

54 "In Hwange National Park . . ." Moss, *Elephant Memories*, pp. 315–16.

55 Bledsoe, *Brown Bear Summer*, pp. 171–76.

56 Lynn Rogers, interview by Susan McCarthy, July 15, 1993.

56 Paul Leyhausen, *Cat Behavior: The Predatory and Social Behavior of Domestic and Wild Cats*, trans. by Barbara A. Tonkin (New York and London: Garland STPM Press, 1979), pp. 286–87.

56 Chadwick, *Beast the Color of Winter*, p. 19.

56 "A peregrine falcon father . . ." Marcy Cottrell Houle, *Wings for My Flight: The Peregrine Falcons of Chimney Rock* (Reading, MA: Addison-Wesley Publishing Co., 1991), p. 105.

56 "One experimenter decided . . ." Harvey A. Hornstein, *Cruelty and Kindness: A New Look at Oppression and Altruism* (Englewood Cliffs, NJ: Prentice-Hall, 1976), p. 81. Citing experiments by Professor Donald O. Hebb.

57 "In this case, however . . ." Herbert S. Terrace, *Nim: A Chimpanzee Who Learned Sign Language* (New York: Washington Square Press, 1979), p. 44. The mother chimpanzee's apprehensions were justified: she was tranquilized and the infant was taken away, named Nim Chimpsky, and taught 125 words of American Sign Language. Years later he was returned to the institute.

57 Hans Kruuk, *The Spotted Hyena: A Study of Predation and Social Behavior* (Chicago: University of Chicago Press, 1972), p. 161.

57 "Pandora, a two-year-old . . ." Chadwick, *Beast the Color of Winter*, p. 26.

58 "In Africa a buffalo . . ." George B. Schaller, *The Serengeti Lion: A Study of Predator-Prey Relations* (Chicago: University of Chicago Press, 1972), p. 266.

58 Kruuk, *Spotted Hyena*, p. 161.

58 "For a nature program . . ." "Cheetahs in the Land of Lions," an episode of *Nature: with George Page*, 1992.

59 Darwin is quoted by Peter J. Bowler, *The Fontana History of the Environmental Sciences* (London: HarperCollins, 1992), pp. 480–81. The sentences before the quote read: "Darwin was led to stress those accounts of animals which depict their behavior as 'almost human.' He made no experiments of his own, and relied upon anecdotal information supplied by hunters, zookeepers and the like." The story of the "heroic" little monkey is found in Darwin's *Descent of Man; and Selections in Relation to Sex*, pp. 89, 95 (Norwalk, CT: Heritage Press Edition, 1972).

59 Moss, *Elephant Memories*, p. 162.

60 "Koko, a gorilla . . ." Francine Patterson and Eugene Linden, *The Education of Koko* (New York: Holt, Rinehart and Winston, 1981), pp. 135–36.

61 "The chimpanzee Viki . . ." R. Allen Gardner and Beatrix T. Gardner, "A Cross-Fostering Laboratory," in *Teaching Sign Language to Chimpanzees*, R. Allen Gardner, Beatrix T. Gardner, and Thomas E. Van Cantfort, eds., 1–28 (Albany: State University of New York Press, 1989), p. 8.

61 "In a remarkable use . . ." Beatrix T. Gardner, Allen Gardner, and Susan G. Nichols, "The Shapes and Uses of Signs in a Cross-Fostering Laboratory," in *Teaching Sign Language to Chimpanzees*, p. 65.

62 "Biologists who arrived . . ." "A Letter from the Field," Luis Baptista, *Pacific Discovery* 16 (4): pp. 44–47.

62 Sherwin Carlquist, *Island Life: A Natural History of the Islands of the World* (Garden City, NY: Natural History Press, 1965), pp. 337–41.

62 "When Washoe grew . . ." Roger S. Fouts, Deborah H. Fouts, and Thomas E. Van Cantfort, "The Infant Loulis Learns Signs from Cross-Fostered Chimpanzees," in *Teaching Sign Language to Chimpanzees*, pp. 280–92. Also personal communication.

63 Ludwig Wittgenstein, *Philosophical Investigations*, 3rd ed., trans. by G. E. M. Anscombe (New York: Macmillan Co., 1968), p. 174.

Chapter 4: Love and Friendship

64 J. H. Williams, *Elephant Bill* (Garden City, NY: Doubleday & Co., 1950), pp. 82–84.

65 "Yet love is not . . ." For example, Carroll E. Izard (*Human Emotions*, New York and London: Plenum Press, 1977) does not include love on his list of the eight basic emotions.

65 Catherine Roberts, *The Scientific Conscience: Reflections on the Modern Biologist and Humanism* (New York: George Braziller, 1967).

65 " 'It's important to remember . . .' " Janine Benyus, *Beastly Behaviors: A Zoo Lover's Companion: What Makes Whales Whistle, Cranes Dance, Pandas Turn Somersaults, and Crocodiles Roar: A Watcher's Guide to How Animals Act and Why* (Reading, MA: Addison-Wesley Publishing Co., 1992), p. 52.

66 Thomas, *Hidden Life of Dogs*.

66 "Thomas has been castigated . . ." Patricia Holt, "Puppy Love Isn't Just For People: Author Says Dogs, Like Humans, Can Bond," *San Francisco Chronicle*, December 9, 1993.

67 "Females of a southeast Asian . . ." I owe this description to Professor Richard I. Vane-Wright. It originally derives from Miriam Rothschild, "Female Butterfly Guarding Eggs," in *Antenna*, London, Vol. 3 (1979), p. 94.

67 J. Traherne Moggridge, *Harvesting Ants and Trap-door Spiders: Notes and Observations on Their Habits and Dwellings* (London: L. Reeve & Co., 1873), pp. 113–14.

69 " 'The female appeared to be . . .' " Bertold P. Wiesner and Norah M. Sheard, *Maternal Behavior in the Rat* (Edinburgh and London: Oliver & Boyd, 1933), pp. 121–22.

69 " 'It is not uncommon . . .' " Tony Gaston and Garry Donaldson, "Banding Thick-billed Murre Chicks," *Pacific Seabirds* 21 (1994), pp. 4–6.

69 "In contrast to . . ." Bill Clark, *High Hills and Wild Goats* (Boston: Little, Brown and Co., 1990), p. 34.

69 "Some biologists suggest . . ." Bettyann Kevles, *Females of the Species: Sex and Survival in the Animal Kingdom* (Cambridge, MA: Harvard University Press, 1986), p. 154.

70 "Biologists studying wild dogs . . ." Frame and Frame, *Swift & Enduring: Cheetahs and Wild Dogs of the Serengeti*, p. 157.

70 "In a typical . . ." Anne Innis Dagg and J. Bristol Foster, *The Giraffe: Its*

Biology, Behavior, and Ecology (New York: Van Nostrand Reinhold Co., 1976), pp. 38–39.

70 " 'When I approached . . .' " Quoted in Faith McNulty, *The Whooping Crane: The Bird That Defies Extinction* (New York: E. P. Dutton & Co., 1966), p. 37.

71 "To the north of . . ." Stanley P. Young, *The Wolves of North America: Their History, Life Habits, Economic Status, and Control* (Part II: "Classification of Wolves" by Edward A. Goldman) (Washington, DC: American Wildlife Institute, 1944), pp. 109–10, citing a 1935 article by Peter Freuchen.

71 "It is estimated that . . ." Devra G. Kleiman and James R. Malcolm, "The Evolution of Male Parental Investment in Mammals," in *Parental Care in Mammals*, David J. Gubernick and Peter H. Klopfer, eds. (New York: Plenum Press, 1981).

71 Gerald Durrell, *Menagerie Manor* (New York: Avon, 1964), pp. 127–29.

72 Macdonald, *Running with the Fox*, pp. 140–42.

72 "Researchers studying wild zebras . . ." Cynthia Moss, *Portraits in the Wild: Behavior Studies of East African Mammals* (Boston: Houghton Mifflin Co., 1975), pp. 104–05.

73 ". . . one adolescent wild baboon . . ." Strum, *Almost Human*, p. 40.

74 "In a captive baboon colony . . ." Rowell, *Social Behaviour of Monkeys*, p. 76.

74 Montgomery, *Walking with the Great Apes*, p. 43.

75 "Researchers in Africa kidnapped . . ." Hans Kummer, *Social Organization of Hamadryas Baboons; A Field Study* (Chicago and London: University of Chicago Press, 1968), p. 63. This study refers to such caretaking by male baboons as mothering and as "maternal" behavior.

75 "The experimenters who gave the mother rat . . ." Wiesner and Sheard, *Maternal Behavior*, p. 148.

75 " 'At Northrepps Hall, near Cromer . . .' " Robert Cochrane, "Some Parrots I Have Known," in *The Animal Story Book*, The Young Folks Library, Vol. IX (Boston: Hall & Locke Co., 1901), pp. 208–09.

76 ". . . a wildebeest calf who . . ." Kruuk, *Spotted Hyena*, p. 171.

76 "One young wild elephant . . ." Moss, *Elephant Memories*, p. 267.

77 "In one pack of wild dogs . . ." Schaller, *Serengeti Lion*, p. 332.

77 Françoise Patenaude, "Care of the Young in a Family of Wild Beavers, *Castor canadensis*," *Acta Zool. Fennica* 174 (1983), pp. 121–22.

78 ". . . a monkey kept alone will work . . ." Rowell, *Social Behaviour of Monkeys*, p. 110.

78 "Elephants appear to make . . ." Moss, *Portraits in the Wild*, pp. 16–17.

79 Hans Kruuk, *The Social Badger; Ecology and Behaviour of a Group-living Carnivore (Meles meles)* (Oxford, England: Oxford University Press, 1989), p. 146.

80 John J. Teal, Jr., "Domesticating the Wild and Woolly Musk Ox," *National Geographic* (June 1970); also Anne Fadiman, "Musk Ox Ruminations," *Life* (May 1985).

80 ". . . a hand-reared leopard was . . ." Singh, *Tiger! Tiger!*, p. 207.

80 Michael P. Ghiglieri, *East of the Mountains of the Moon: Chimpanzee Society in the African Rain Forest* (New York: Free Press/Macmillan, 1988), p. 119.

80 "A group of wild dogs . . ." Frame and Frame, *Swift & Enduring*, pp. 85–88.

81 "In Madagascar a brown lemur . . ." Alison Jolly, *Lemur Behavior: A Madagascar Field Study* (Chicago: University of Chicago Press, 1966), pp. 123, 126–28.

81 Leyhausen, *Cat Behavior*, pp. 242–43.

82 "Wild beavers, given time . . ." Hope Ryden, *Lily Pond: Four Years with a Family of Beavers* (New York: William Morrow & Co, 1989).

82 "Lucy, a chimpanzee . . ." E. S. Savage, Jane Temerlin, and W. B. Lemmon, "The Appearance of Mothering Behavior Toward a Kitten by a Human-Reared Chimpanzee," paper delivered at the Fifth Congress of Primatology, Nagoya, Japan, 1974.

82 "It is also reported . . ." Chadwick, *The Fate of the Elephant*, pp. 270–71.

83 Professor William Jankowiak, interview by Susan McCarthy, December 15, 1992. Also see Daniel Goleman, "Anthropology Goes Looking for Love in All the Old Places," *New York Times*, November 24, 1992.

83 Professor Charles Lindholm, interview by Susan McCarthy, January 12, 1993.

84 Jane Goodall, *In the Shadow of Man*, revised edition (Boston: Houghton Mifflin Company, 1988), p. 194.

84 "The butterfly fish of . . ." John P. Hoover, *Hawaii's Fishes: A Guide for Snorkelers, Divers and Aquarists* (Honolulu: Mutual Publishing, 1993), pp. 26–27.

85 A. J. Magoun & P. Valkenburg, "Breeding Behavior of Free-ranging Wolverines *(Gulo)" Acta Zool. Fennica* 174 (1983), pp. 175–77.

85 " '*After all* . . .' " Edna St. Vincent Millay, "Passer Mortuus Est," in *Collected Lyrics* (New York: Washington Square Press, 1959), p. 56.

86 "Konrad Lorenz said that . . ." Pryor, *Lads Before the Wind*, p. 171.

86 Mattie Sue Athan, *Guide to a Well-Behaved Parrot* (Hauppauge, NY: Barron's Educational Series, 1993), p. 138.

87 " 'It sickens me when people . . .' " David Cantor, "Items of Property," pp. 280–90 in *The Great Ape Project*. In August 1994 a spokesperson for the Cleveland Metropark Zoo said that Timmy had been sent to the Bronx Zoo, where he had fathered four baby gorillas. Katie was sent to the Fort Worth Zoo to serve as an aunt to other baby gorillas.

87 "Much has been made of the significant rates of infidelity . . ." For an overview see Natalie Angier, "Mating for Life? It's Not for the Birds or the Bees," *New York Times*, August 21, 1990.

87 ". . . male prairie voles who have formed . . ." James T. Winslow, Nick Hastings, C. Sue Carter, Carroll R. Harbaugh, and Thomas R. Insel, "A Role for Central Vasopressin in Pair Bonding in Monogamous Prairie Voles," *Nature* 365 (7 October 1993), pp. 545–48.

87 Moss, *Elephant Memories*, pp. 100–01.

88 "The male she had paired with . . ." Moss, *Portraits in the Wild*, p. 49.

88 Ryden, *God's Dog*, pp. 60–62.

89 George Archibald, "Gee Whiz! ICF Hatches a Whooper," *The ICF Bugle* (July 1982). In a letter to Jeffrey Masson of July 25, 1994, George Archibald

adds the following interesting details: "After she laid her egg in 1982, the egg was replaced with a dummy egg, and I spent the night in my shack beside Tex's nest. My duty was to protect Tex from predators, and we hoped that allowing Tex to incubate the first egg would stimulate her to produce a second egg. About midnight a downpour of rain accompanied by strong winds descended on the Baraboo Hills. Tex was drenched as she sat on her nest. Every few minutes she emitted low frequency Contact Calls (low purring sound), and I answered. If I called to her, she immediately responded with a Contact Call. When the radio announced a tornado warning, I left the shack and with thunder crashing and lightning flashing, I picked up Tex, held her under my arm, and walked down the hay field to her shed. I talked and she Contact Called all the way there. During this emergency, I felt a strong emotional connection with Tex."

89 Gavin Maxwell, *Raven, Seek Thy Brother* (London: Penguin Books, 1968), pp. 59–61.

Chapter 5: Grief, Sadness, and the Bones of Elephants

91 Houle, *Wings for My Flight*, pp. 75–87. The female peregrine had reportedly been shot. The two surviving nestlings fledged successfully.

92 "According to naturalist Georg Steller . . ." Quoted in H. C. Bernhard Grzimek, ed., *Grzimek's Animal Life Encyclopedia*, Vol. 12 (New York: Van Nostrand Reinhold Co., 1975).

93 Thomas, *Hidden Life of Dogs*.

94 "Ackman and Alle, two circus horses . . ." Henderson, *Circus Doctor*, p. 78.

94 Pryor, *Lads Before the Wind*, pp. 276–77.

94 "Researchers who had caught . . ." Antony Alpers, *Dolphins: The Myth and the Mammal* (Boston: Houghton Mifflin Co., 1960), pp. 104–05.

95 "Lions do not form . . ." Thomas, "Reflections: The Old Way," p. 91.

95 Moss, *Elephant Memories*, pp. 269–71.

95 ". . . African elephants surrounding a dying matriarch . . ." Moss, *Portraits in the Wild*, p. 34.

95 Moss, *Elephant Memories*, pp. 272–73.

96 "Three small groups of chimpanzees . . ." Geza Teleki, "Group Response to the Accidental Death of a Chimpanzee in Gombe National Park, Tanzania," *Folia Primatol* 20 (1973), pp. 81–94.

97 "A chimpanzee at the Arnhem Zoo . . ." De Waal, *Chimpanzee Politics*, pp. 67–70.

97 " 'if they do not get companionship . . .' " Lars Wilsson, *My Beaver Colony*, trans. by Joan Bulman (Garden City, NY: Doubleday & Co., 1968), pp. 61–62.

98 ". . . 'bull areas.' " Moss, *Elephant Memories*, p. 112.

98 Leyhausen, *Cat Behavior*, pp. 287–88.

99 Goodall, *Through a Window: My Thirty Years with the Chimpanzees of Gombe* (Boston: Houghton Mifflin Co., 1990), p. 230.

100 " 'Hum-Hum had lost all joy . . .' " Cited in Robert M. Yerkes and Ada W.

Yerkes, *The Great Apes: A Study of Anthropoid Life* (New Haven, CT: Yale University Press, 1929), p. 472.

100 "A pilot whale celebrity . . ." Pryor, *Lads Before the Wind*, pp. 82–83.

100 "Yet at Sea World, in San Diego . . ." Robert Reinhold, "At Sea World, Stress Tests Whale and Man," *New York Times*, April 4, 1988, p. A9.

100 ". . . 'just moped to death.'" Pryor, *Lads Before the Wind*, p. 132.

101 "'It seems reasonable to allow . . .'" McFarland, ed., *Oxford Companion to Animal Behavior*, p. 599.

102 ". . . 'placed alone in the "depression chamber" . . .'" Harlow said that his "device was designed on an intuitive basis to reproduce such a well [of despair] both physically and psychologically for monkey subjects." See the trenchant criticism by James Rachels in "Do Animals Have a Right to Liberty?" in *Animal Rights and Human Obligations*, Tom Regan and Peter Singer, eds. (Englewood Cliffs, NJ: Prentice-Hall, 1976), p. 211. See, too, Peter Singer's criticism in Chapter 2 of his *Animal Liberation*.

102 "Even when months had passed . . ." "Do Animals Have a Right to Liberty?" in Regan and Singer, eds., *Animal Rights*, p. 211. See, too, the fine criticism of Harlow's work in Chapter 2 of Peter Singer's influential *Animal Liberation* (New York Review, 1975); the original article by Harlow is written with Stephen J. Suomi: "Depressive Behavior in Young Monkeys Subjected to Vertical Chamber Confinement," *Journal of Comparative and Physiological Psychology* 80 (1972), pp. 11–18. Harlow published his articles in prestigious journals. For example, see his "Love in Infant Monkeys," *Scientific American* 200 (1959), pp. 68–74; and "The Nature of Love," *American Psychologist*, 13 (1958), pp. 673–85. A useful general critique is found in *Psychology Experiments on Animals: A Critique of Animal Models of Human Psychopathology* by Brandon Kuker-Reines for the New England Anti-Vivisection Society, 1982, who remarks in a telling aside that "the apparent quest to reveal the 'true man' through monkey experimentation is symptomatic of an identity crisis rather than scientific progress" (p. 68).

102 ". . . 'learned helplessness.'" Martin E. P. Seligman, *Helplessness: On Depression, Development, and Death* (San Francisco, CA: W. H. Freeman & Co., 1975), pp. 23–25. While restrained, each dog was given sixty-four shocks of 6.0 milliamperes, lasting for five seconds.

103 ". . . talking to battered women . . ." Lenore Walker has powerfully delineated the role of learned helplessness in the lives of battered women. See her *Terrifying Love: Why Battered Women Kill and How Society Responds* (New York: Harper & Row, 1989).

103 "One experimenter raised rhesus monkeys in solitude . . ." J. B. Sidowski, "Psychopathological Consequences of Induced Social Helplessness During Infancy," in *Experimental Psychopathology: Recent Research and Theory*, H. D. Kimmel, ed. (New York: Academic Press, 1971), pp. 231–48.

104 ". . . a special mourning howl . . ." Russell J. Rutter and Douglas H. Pimlott, *The World of the Wolf* (Philadelphia and New York: J. B. Lippincott Co., 1968), p. 138; Lois Crisler, *Captive Wild* (New York: Harper & Row, 1968), p. 210.

104 "When Marchessa . . ." Ian Redmond, "The Death of Digit," *International Primate Protection League Newsletter* 15, No. 3, December 1988, p. 7.

104 " 'In disappointment the young specimen. . . ." Yerkes and Yerkes, *Great Apes*, p. 161.

105 "Emotional tears are different . . ." William Frey, II, *Crying: The Mystery of Tears*, with Muriel Langseth (Minneapolis: Winston Press, 1985). Emotional tears are also called psychogenic tears. It is unclear where tears of pain fit into these categories.

105 ". . . the one body product that may . . ." S. B. Ortner, "Shera purity," *American Anthropologist* 75 (1973), pp. 49–63. Quoted in Paul Rozin and April Fallon, "A Perspective on Disgust," *Psychological Review* 94 (1987), pp. 23–41.

105 ". . . Nim Chimpsky . . ." Terrace, *Nim*, p. 56.

105 "Tears have been seen . . ." De Grahl, *Grey Parrot*, p. 189.

105 ". . . especially apt to have tears . . ." Victor B. Scheffer, *Seals, Sea Lions, and Walruses: A Review of the Pinnipedia* (Stanford, CA: Stanford University Press, 1958), p. 22. Also Frey, *Crying*.

106 *Macacus maurus*, the Celebes macaque, is now denoted *Cynomacaca maurus*. An influential German book in its time, Karl Friedrich Burdach's *Blicke ins Leben* (3 Vols., Leipzig, Germany: Leopold Woss, 1842), Vol. 2, p. 130, cites examples of female seals who "shed copious tears when they were abused," giraffes who cried when they were removed from their companions, and tears in fur seals when their young were stolen *(geraubt)* and in an elephant seal when it was treated roughly.

106 Frey, *Crying*, p. 141.

106 "Tears fell from the eyes . . ." Volker Arzt and Immanuel Birmelin, *Haben Tieren ein Bewusstsein?: Wenn Affen lügen, wenn Katzen denken und Elefanten traurig sind* (Munich: C. Bertelsmann, 1993), p. 154.

106 R. Gordon Cummings, *Five Years of a Hunter's Life in the Far Interior of South Africa* (1850), quoted in Richard Carrington, *Elephants: A Short Account of Their Natural History, Evolution and Influence on Mankind* (London: Chatto & Windus, 1958), pp. 154–55.

107 George Lewis, as told to Byron Fish, *Elephant Tramp* (Boston: Little, Brown and Co., 1955), pp. 52, 188–89.

107 Victor Hugo, *Carnet intime 1870–1871*, publié et presenté par Henri Guillemin. (Paris: Gallimard, 7th ed., 1953), p. 88.

108 "shedding tears when scolded . . ." Chadwick, *Fate of the Elephant*, p. 327.

108 "Observing young orphaned . . ." Chadwick, *Fate of the Elephant*, p. 384.

108 "Perhaps the position somehow prevents drainage . . ." This suggestion was proposed by Dr. William Frey.

109 ". . . beavers also weep copiously . . ." L. S. Lavrov, "Evolutionary Development of the Genus *Castor* and Taxonomy of the Contemporary Beavers of Eurasia," *Acta Zool. Fennica* 174 (1983), pp. 87–90.

109 Dian Fossey, *Gorillas in the Mist* (Boston: Houghton Mifflin Co., 1983), p. 110.

109 D. M. Frame, trans., *The Complete Works of Montaigne, Vol. 2* (Garden City, NY: Anchor Books, 1960), pp. 105–09.

Chapter 6: A Capacity for Joy

111 "It *knew* it was free . . ." Kenneth S. Norris, *Dolphin Days: The Life and Times of the Spinner Dolphin* (New York: W. W. Norton & Co., 1991), pp. 129–30.

111 ". . . 'what obtains after . . .' " Izard, *Human Emotions*, pp. 239–45.

112 "Lions purr . . ." Schaller, *Serengeti Lion*, pp. 104, 304.

112 "Happy gorillas are said to sing." Reported in Montgomery, *Walking with the Great Apes*, p. 146.

112 "Howling wolves . . ." Thomas, *Hidden Life of Dogs*, p. 40.

113 Lynn Rogers, interview by Susan McCarthy, July 15, 1993.

113 Darwin to Susan Darwin, 1838, *The Correspondence of Charles Darwin Volume 2; 1837–1843* (Cambridge, England: Cambridge University Press, 1986).

113 Norris, *Dolphin Days*, pp. 42–43.

114 "When the ice finally melted . . ." Ryden, *Lily Pond*, p. 104.

114 "Nim Chimpsky . . ." Terrace, *Nim*, p. 412.

114 ". . . 'gorilla hug.' " Patterson and Linden, *The Education of Koko*, p. 185.

114 Carolyn A. Ristau and Donald Robbins, "Language in the Great Apes: A Critical Review," *Advances in the Study of Behavior*, Vol. 12, 141–255 (1982), p. 229.

114 ". . . 'singing in the rain.' " Roger Fouts, interview by Susan McCarthy, December 10, 1993.

115 ". . . some goats have been seen . . ." Chadwick, *Beast the Color of Winter*, pp. 150–51.

115 Jane Goodall and David A. Hamburg, "Chimpanzee Behavior as a Model for the Behavior of Early Man," in Silvano Arieti, ed., *American Handbook of Psychiatry*, 2nd ed. (New York: Basic Books, 1975), pp. 20–27. Cited here from Carl N. Degler *In Search of Human Nature: The Decline and Revival of Darwinism in American Social Thought* (New York: Oxford University Press, 1991), p. 336.

115 Terrace, *Nim*, pp. 140–42.

115 "Two male bottle-nosed dolphins . . ." Alpers, *Dolphins*, p. 102.

116 Moss, *Elephant Memories*, pp. 124–25.

116 Wilsson, *My Beaver Colony*, pp. 92–93.

117 "The presence of such traits . . ." Richard Monastersky, "Boom in 'Cute' Baby Dinosaur Discoveries," *Science News* 134 (October 22, 1988), p. 261.

117 "When a young sparrow . . ." Arzt and Birmelin, *Haben Tieren ein Bewusst-sein?*, p. 173.

117 Wilsson, p. 131.

118 ". . . beaver's 'subjective feelings . . .' " Ryden, *Lily Pond*, pp. 185–87.

118 "That a tiger is condemned . . ." On the other hand, Elizabeth Marshall Thomas is of the opinion that "the best life for a large captive cat is that of a circus performer. A fortunate circus tiger, in my view, might share a cage with another, compatible tiger, in a collection of ten or twenty fellow tigers whose owners not only train them and perform with them but in all ways share their lives." *The Tribe of Tiger: Cats and Their Culture* (New York: Simon & Schuster, 1994), p. 194.

118 Gebel-Williams with Reinhold, *Untamed*, p. 310.

119 Karen Pryor and Kenneth S. Norris, eds. *Dolphin Societies: Discoveries and Puzzles* (Berkeley: University of California Press, 1991), p. 346.

119 "Horse trainers commonly . . ." Heywood Hale Broun, "Ever Indomitable, Secretariat Thunders Across the Ages," *New York Times*, May 30, 1993, p. 23.

119 Ralph Dennard, interview by Jeffrey Moussaieff Masson and Susan McCarthy, September 24, 1993.

120 "Washoe may have had . . ." Roger Fouts, interview by Susan McCarthy, December 10, 1993.

121 "It looks as if . . ." De Waal, *Chimpanzee Politics*, p. 26.

121 ". . . 'exploded with joy' . . ." George B. Schaller, *The Last Panda* (Chicago and London: University of Chicago Press, 1993), p. 66.

121 J. Lee Kavanau, "Behavior of Captive White-footed Mice," *Science* 155 (March 31, 1967): pp. 1623–39.

122 ". . . 'lamp-pulling and squirting behavior . . .' " P. B. Dews, "Some Observations on an Operant in the Octopus," *Journal of the Experimental Analysis of Behavior* 2 (1959): pp. 57–63. Reprinted in Thomas E. McGill, ed., *Readings in Animal Behavior* (New York: Holt, Rinehart and Winston, 1965).

122 F. Fraser Darling, *A Herd of Red Deer: A Study in Animal Behavior* (London: Oxford University Press, 1937), p. 35.

123 "When Indah . . ." "Orangutan Escapes Exhibit, Mingles with Zoo Visitors," *San Francisco Chronicle*, June 19, 1993 (Associated Press story).

124 "Play . . . has been increasingly studied . . ." See M. Bekoff and J. A. Byers: "A Critical Reanalysis of the Ontogeny and Phylogeny of Mammalian Social and Locomotor Play: An Ethological Hornet's Nest." In K. Immelmann et al., *Behavioral Development: The Bielefeld Interdisciplinary Project.* (Cambridge, England: Cambridge University Press, 1981), pp. 296–337.

124 Robert Fagen, *Animal Play Behavior* (New York, Oxford: Oxford University Press, 1981), pp. 3–4.

124 "Biologists continue to be dismayed . . ." *Ibid.*, pp. 17–18. Fagen notes that whenever he lectured on animal play, "Afterwards, to my discomfort and embarrassment, I would chiefly be asked 'people questions.' "

124 " '. . . this behavior fascinates . . .' " Fagen, p. 494.

124 Robert A. Hinde, *Animal Behavior* (New York: McGraw-Hill, 1966).

124 Marc Bekoff, "Kin Recognition and Kin Discrimination" (letter) *Trends in Ecology and Evolution* 7 (3), March 1992, p. 100.

125 Moss, *Elephant Memories*, pp. 85, 142–43, 171.

125 Kruuk, *Spotted Hyena*, pp. 249–50.

125 ". . . Norma, a young elephant . . ." Lewis, *Elephant Tramp*, pp. 128–29.

126 Terrace, *Nim*, pp. 228–29.

126 "Alaskan buffalo . . . playing on ice." Gary Paulsen, *Winterdance: The Fine Madness of Running the Iditarod* (New York: Harcourt, Brace & Co., 1994), p. 193.

126 "Two grizzly bears . . ." Chadwick, *Beast the Color of Winter*, p. 70.

126 "Tiger cubs and leopards . . ." Singh, *Tiger! Tiger!*, pp. 72–73.

126 "The bonobo covers its eyes . . ." De Waal, *Peacemaking*, p. 195.

126 ". . . domes of the Kremlin . . ." Jeffery Boswall, "Russia Is for the Birds," *Discover* (March 1987), p. 78.

127 ". . . Komodo dragon . . . played with a shovel . . ." Craven Hill, "Playtime at the Zoo," *Zoo-Life* 1: pp. 24–26.

127 ". . . alligator in Georgia . . ." James D. Lazell, Jr., and Numi C. Spitzer, "Apparent Play Behavior in an American Alligator," *Copeia* (1977): p. 188.

127 ". . . Koko . . . pretends to brush her teeth . . ." Patterson and Linden, *Education of Koko*, picture caption.

127 ". . . 'That's a hat.' " Roger Fouts, interview by Susan McCarthy, December 10, 1993.

127 ". . . dolphins vied for . . ." Alpers, *Dolphins*, pp. 90–93.

127 ". . . play similar keep-away . . ." Norris, *Dolphin Days*, pp. 259–60.

127 "Beluga whales carry stones . . ." Fred Bruemmer, *"White Whales on Holiday,"* *Natural History* (January 1986): pp. 40–49.

127 "Lions, both adults and cubs . . ." Schaller, *Serengeti Lion*, pp. 163–64.

127 ". . . dolphin teased a fish . . ." Alpers, *Dolphins*, p. 90.

127 "Ravens tease peregrines . . ." Houle, *Wings for My Flight*, p. 23.

128 ". . . crows may pull their tails . . ." Crumley, *Waters of the Wild Swan*, pp. 53–54.

128 ". . . hyenas . . . catching and killing such a fox . . ." Macdonald, *Running with the Fox*, pp. 78–79.

128 "Sifaka lemurs . . ." Jolly, *Lemur Behavior*, p. 59.

128 ". . . cricket was taught to elephants . . ." Carrington, *Elephants*, pp. 216–17.

129 ". . . dolphins . . . foul play . . ." Pryor, *Lads Before the Wind*, pp. 66–67.

129 "The kangaroos preferred to wrestle and box . . ." Geoffrey Morey, *The Lincoln Kangaroos* (Philadelphia: Chilton Books, 1963), pp. 53–60.

130 "Tatu . . ." Rasa, *Mongoose Watch*, pp. 44–45, 142–44.

130 ". . . beavers and otters were present . . ." Hope Sawyer Buyukmihci, *The Hour of the Beaver* (Chicago: Rand McNally & Co., 1971), pp. 97–98.

130 ". . . mangabey and red-tailed monkeys . . ." Ghiglieri, *East of the Mountains of the Moon*, p. 26.

130 Chadwick, *Fate of the Elephant*, pp. 423–24.

131 Bert Hölldobler and Edward O. Wilson, *The Ants* (Cambridge, MA: Belknap Press/Harvard University Press, 1990), p. 370.

131 Henry Walter Bates, *The Naturalist on the River Amazons: A Record of Adventures, Habits of Animals, Sketches of Brazilian and Indian Life, and Aspects of Nature under the Equator, During Eleven Years of Travel* (New York: Humboldt Publishing Co., 1863), pp. 259–60.

Chapter 7: Rage, Dominance, and Cruelty in Peace and War

133 ". . . Cosimo de' Medici shut a giraffe . . ." Dagg and Foster, *The Giraffe*, p. 3.

133 "While aggression among animals is a favored topic . . ." Konrad Lorenz, one of the founders of ethology, wrote a celebrated book on aggression.

Ruth Klüger, in her *weiter leben: Eine Jugend* (Göttingen, Germany: Wall-stein, 1992, p. 186), in discussing programmed and flexible learning behavior, makes this acid observation: "On the other hand, one cannot predict the behavior of the behavioral researcher: he was a Nazi and became a great professor at that time, and then he once again became a sensible contemporary with justifiable political views. Naturally evil remained for him always only the 'so-called evil,' and the temptation to evil, which lies in human freedom, he chose not to acknowledge. He confused it stubbornly with the preprogrammed animal aggression, which he had so thoroughly researched."

134 ". . . relations may not be hierarchical." See, for example, Ryden's *God's Dog*, p. 223, where she complains ". . . my animals got along so well that I was unable to determine their relative ranks."

134 "Anger and other emotions related . . ." What animal aggression says about human aggression is debated. Richard Lewontin writes that "there is in fact not a shred of evidence that the anatomical, physiological, and genetic basis of what is called aggression in rats has anything in common with the German invasion of Poland in 1939." (*Biology as Ideology: The Doctrine of DNA*. New York: HarperCollins, 1991, p. 96.) On the other hand, the Renaissance historian Richard Trexler said that "without animal behavior studies, I would understand much less about human aggression in Italy in the fourteenth century." [Personal communication.]

135 " 'Animals fight . . .' " *Aussichten auf den Bürgerkrieg*. Frankfurt am M., 1993.

135 ". . . Kasakela apes . . ." Jane Goodall, *The Chimpanzees of Gombe; Patterns of Behavior* (Cambridge, MA and London: Harvard University Press, 1986), p. 502.

136 "Bands of dwarf mongooses . . ." Rasa, *Mongoose Watch*, pp. 230–31.

136 Kruuk, *Spotted Hyena*, pp. 254–56.

137 "Parrots have been known . . ." Mattie Sue Athan, *Guide to a Well-Behaved Parrot*, p. 138.

138 ". . . dominance *relationships* . . . dominance *ranks* . . ." Irwin S. Bernstein, "Dominance: the Baby and the Bathwater," *Behavioral and Brain Sciences* 4 (1981), pp. 419–29. (Followed by peer commentary.)

138 ". . . a female and her adolescent daughter." Thelma Rowell has suggested that rank relationships may be better characterized as subordinance rather than dominance relationships, since it is the giving way by one animal that constitutes a decision not to fight. In her view a dominance rank does not express the social character of a baboon, but is what is "left over" after his degree of subordination is accounted for. (Rowell, *Social Behaviour of Monkeys*, pp. 162–63)   Consider ring-tailed lemurs, a species in which females dominate males. Males seem to have a clear dominance order among themselves, and females a less apparent one. Researcher Alison Jolly noted, "Females . . . were far less 'status-conscious.' They might gratuitously chase each other or the males and would cuff any animal which came too close. However, not only did they spit less frequently, but they did not carry themselves in a particularly erect or cringing posture, nor did they keep an eye on dominant troop members and dodge their approach. The general dominance

of females over males seems to rise out of the same attitudes: an insouciant female would cuff any animal, but a male was subordinate to any animal it could not bully." *(Lemur Behavior: A Madagascar Field Study.* Alison Jolly. (Chicago: University of Chicago Press, 1966), pp. 104–07.) Also Alison F. Richard, "Malagasy Prosimians: Female Dominance," in *Primate Societies,* eds. Barbara B. Smuts et al., pp. 25–33 (Chicago and London: University of Chicago Press, 1986).

139 "In the hamadryas baboon . . ." Christian Bachmann and Hans Kummer, "Male Assessment of Female Choice in Hamadryas Baboons," *Behavioral Ecology and Sociobiology* 6 (1980), pp. 315–21. This paper continues a tradition of referring to male hamadryas baboons as "owners" of females.

139 Strum, *Almost Human,* pp. 118–20.

139 Leyhausen, *Cat Behavior,* pp. 256–57.

139 "This may be why . . ." Lemurs are by no means the only species that exhibit female dominance. The recently rediscovered mountain pygmy-possum (Burramys parvus) is a mouse-sized marsupial, previously known only from fossils. Living high in the Australian Alps, they must survive fierce winters. They are thought to show an unusual form of female dominance. The females occupy good foraging habitat year-round and in the winter hibernate in nests with their daughters. The males, who mate with many females and who do not take care of the young, move into these areas in the summer. In the winter, apparently ousted by the females, they move to poorer habitats where they hibernate alone or with other males. Fewer males survive the winter, so although equal numbers of male and female pygmy-possums are born, adult males are much less common than females. *(The Mountain Pygmy-possum of the Australian Alps.* Ian Mansergh and Linda Broome. Kensington, NSW, Australia: New South Wales University Press, 1994.)

139 "Scientists' behavior may . . ." President Theodore Roosevelt, "an enthusiastic imperialist and a staunch believer in the superiority of the Anglo-Saxon race, was also a renowned Great White Hunter who devoted much of his life to killing large animals throughout the world and writing books recounting his adventures." The early Canadian conservationists John Muir (founder of the Sierra Club) and William J. Long engaged the president in a much followed debate in the popular press. When Roosevelt contended that they lacked manliness and did not know "the heart of the wild thing," Long snapped back with a famous counterattack:

> Who is he to write, "I don't believe for a minute that some of these nature writers know the heart of a wild thing." As to that, I find after carefully reading two of his big books that every time Mr. Roosevelt gets near the heart of a wild thing he invariably puts a bullet through it.

This quote and the earlier one about Roosevelt come from Matt Cartmill's *A View to a Death in the Morning,* pp. 153–54.

140 ". . . scimitar-horned oryx . . ." Clark, *High Hills and Wild Goats*, pp. 67–68.

140 ". . . rape . . . in coatimundis . . ." Bil Gilbert, *Chulo* (New York: Alfred A. Knopf, 1973), pp. 230–31.

141 ". . . white-fronted bee-eaters . . ." S. T. Emlen and P. H. Wrege, "Forced Copulations and Intraspecific Parasitism: Two Costs of Social Living in the White-fronted Bee-eater," *Ethology* 71 (1986), pp. 2–29.

141 ". . . males try to pile on." Robert O. Bailey, Norman R. Seymour, and Gary R. Stewart, "Rape Behavior in Blue-winged Teal," *Auk* 95 (1978), pp. 188–90. Also, David P. Barash, "Sociobiology of Rape in Mallards *(Anas platyrhynchos):* Responses of the Mated Male," *Science* 197 (August 19, 1977), pp. 788–89.

141 ". . . rape the newcomer . . ." Pryor, *Lads Before the Wind*, pp. 78–79.

141 ". . . in the wild . . . dolphins . . ." Natalie Angier, "Dolphin Courtship: Brutal, Cunning and Complex," *New York Times*, February 18, 1992.

141 Kruuk, *Spotted Hyena*, p. 232.

142 ". . . penguins may push one of their . . ." John Alcock, *Animal Behavior: An Evolutionary Approach*, 4th ed. (Sunderland, MA: Sinauer Associates, 1989), pp. 372–73.

142 ". . . giraffe got off the road . . ." Dagg and Foster, *Giraffe*, pp. 36–37.

142 ". . . both appear irritated . . ." Pryor, *Lads Before the Wind*, p. 123.

142 ". . . a young false killer whale . . ." *Ibid.*, p. 214.

143 ". . . colleagues of Pavlov tried to . . ." Quoted in Thomas M. French, *The Integration of Behavior, Volume 1: Basic Postulates* (Chicago: University of Chicago Press, 1952), pp. 156–57.

144 ". . . 'literally eaten alive . . .' " "What Everyone Who Enjoys Wildlife Should Know," pamphlet from Abundant Wildlife Society of North America, Gillette, Wyoming. Also *Abundant Wildlife*, Special Wolf Issue, 1992.

144 "The whistling dog . . ." Michael W. Fox, *The Whistling Hunters: Field Studies of the Asiatic Wild Dog (Cuon Alpinus)* (Albany: State University of New York Press, 1984), p. 63.

145 "A leopard . . . play with captured jackals . . ." Moss, *Portraits in the Wild*, p. 296.

145 Leyhausen, *Cat Behavior*, pp. 128–30.

145 ". . . tiger catches prey . . ." *Ibid.*, pp. 136–37.

145 ". . . lioness has been seen . . ." Thomas, "Reflections: The Old Way," p. 93.

146 ". . . cats . . . play with paper balls . . ." Leyhausen, *Cat Behavior*, p. 137.

147 "Bears, confronted by a river full . . ." Bledsoe, *Brown Bear Summer*, p. 67.

147 "Hyenas invade a flock . . ." Kruuk, *Spotted Hyena*, p. 89.

147 "Such surplus killers . . ." See, for example, Troy R. Mader, "Wolves and Hunting," *Abundant Wildlife*, Special Wolf Issue (1992), p. 3. Accounts of wolves surplus-killing deer in Minnesota, caribou calves in Canada, and Dall sheep in Alaska are used to argue that wolf numbers must be limited. Also photo caption, p. 1.

147 "Both wild and captive hyenas . . ." Kruuk, *Spotted Hyena*, p. 119. Also Stephen E. Glickman, pers. comm., November 5, 1992.

147 ". . . eat some of the surplus." *Ibid.*, pp. 165, 204.

147 "They may not estimate closely . . ." Gerard Gormley, *Orcas of the Gulf; a Natural History* (San Francisco: Sierra Club Books, 1990), p. 85.

148 Schaller, *Serengeti Lion*, p. 383. He adds that lions tend to treat humans as fellow predators rather than as prey.

149 "Congo, a chimpanzee . . ." Desmond Morris, *Animal Days* (New York: Perigord Press/William Morrow & Co., 1980), pp. 222–23.

149 Leyhausen, *Cat Behavior*, pp. 234–35.

150 William Jordan, *Divorce Among the Gulls: An Uncommon Look at Human Nature* (San Francisco: North Point Press, 1991), p. 30.

150 ". . . Bimbo . . . Tabu . . . Mkuba . . ." Sigvard Berggren, *Berggren's Beasts*, translated from the Swedish by Ian Rodger (New York: Paul S. Eriksson, 1970), p. 76.

150 Terrace, *Nim*, pp. 51–52.

151 "The student may be thought of as a model . . ." This is an example of observational learning, and animals have frequently been said to be unable to do this. However, observational learning has been experimentally demonstrated in animals as diverse as cats and octopuses.

151 Irene Pepperberg, interview by Susan McCarthy, February 22, 1993.

151 "A tame parrot may suddenly . . ." De Grahl, *Grey Parrot*, p. 46.

151 Athan, *Guide to a Well-Behaved Parrot*, p. 11.

151 "When Nepo . . . Kianu . . ." Don C. Reed, *Notes from an Underwater Zoo* (New York: Dial Press, 1981), pp. 248–51. Kianu was separated from the other orcas, became visibly depressed, and was sold to a Japanese oceanarium. Nepo died in 1980. Yaka is still at the original oceanarium.

152 De Waal, *Chimpanzee Politics*, p. 168.

152 ". . . in the Arnhem Zoo . . ." *Ibid.*, p. 116. Also De Waal, *Peacemaking Among Primates*.

152 De Waal, *Peacemaking Among Primates*, p. 5.

Chapter 8: Compassion, Rescue, and the Altruism Debate

154 "One evening . . ." Ralph Helfer, *The Beauty of the Beasts: Tales of Hollywood's Animal Stars* (Los Angeles: Jeremy P. Tarcher, 1990), pp. 109–10. Also interview by Susan McCarthy, November 11, 1993.

155 "In Aberdare National Park . . ." Esmond and Chrysse Bradley Martin, *Run Rhino Run* (London: Chatto and Windus, 1982), p. 28.

155 "Young white oryx . . ." Clark, *High Hills and Wild Goats*, p. 198.

156 "A mother Thomson's . . ." Kruuk, *Spotted Hyena*, p. 193.

156 "Thus a researcher . . ." Moss, *Portraits in the Wild*, p. 72.

156 "When a group . . ." Jane Goodall, *With Love* (Ridgefield, CT: Jane Goodall Institute, 1994).

156 "Zebras energetically . . ." Moss, *Portraits in the Wild*, pp. 111–12.

156 "African buffalo . . ." Cited in Schaller, *Serengeti Lion*, p. 262.

156 Bates, *Naturalist on the River Amazons*, pp. 251–52.

157 "Consider the following . . ." Herbert Friedmann, "The Instinctive Emotional Life of Birds," *Psychoanalytic Review* 21 (1934), p. 255. The author, who was the curator of birds at the Smithsonian, delivered this lecture to the Washington Society for Nervous and Mental Diseases. Not only did the author consider the birds too lowly evolved to have compassion, but men were so highly evolved that when they hunted these parrots, "the satisfaction in inflicting cruelty may be fundamentally similar to the pleasure in other forms of endeavor and achievement" (p. 257).

158 Macdonald, *Running with the Fox*, p. 220.

158 "Tatu, a dwarf . . ." Rasa, *Mongoose Watch*, pp. 257–58.

159 "In a case of . . ." *Gorilla: Journal of the Gorilla Foundation* 15 (June 1992), No. 2, p. 8.

159 "Male elephants . . ." Chadwick, *Fate of the Elephant*, p. 94.

159 "An adult pilot whale . . ." Richard C. Connor and Kenneth S. Norris, "Are Dolphins Reciprocal Altruists?" *The American Naturalist* 119, No. 3 (March 1982), p. 363.

160 "Shooting lions . . ." Schaller, *Serengeti Lion*, pp. 25–26.

160 Ralph Dennard, interview by Jeffrey Moussaieff Masson and Susan McCarthy, September 24, 1993.

160 "One family got . . ." Cindy Ott-Bales, interview by Susan McCarthy, September 30, 1993. The baby suffered no ill effects from his single choking episode. Gilly, a Border collie, is trained to notify Ms. Ott-Bales's husband of doorbells and so on.

161 "Another signal dog . . ." Paul Ogden, *Chelsea: The Story of a Signal Dog* (Boston: Little, Brown and Co., 1992), p. 145.

161 Kearton, 1925, cited in Yerkes and Yerkes, *The Great Apes*, p. 298.

161 "An adult wild chimpanzee . . ." Goodall, *With Love*.

161 "It has already been . . ." Terrace, *Nim*, pp. 56–57.

162 ". . . thirty-sixth word . . ." *Ibid.*, 406.

162 "In one grim . . ." Jules H. Masserman, Stanley Wechkin, and William Terris, " 'Altruistic' Behavior in Rhesus Monkeys," *American Journal of Psychiatry* 121 (1964), pp. 584–85.

163 Richard Dawkins, *The Selfish Gene* (New York and Cambridge, England: Cambridge University Press, 1976), p. 14.

163 ". . . Dawkins specifies that he uses the term . . ." *Ibid.*, p. 4.

163 "An example of chimpanzee . . ." Montgomery, *Walking with the Great Apes*, pp. 265–66.

163 "Sniff's mother . . ." Goodall, *Through a Window*, pp. 107–08.

164 ". . . clump of mushrooms . . ." Dawkins, *The Selfish Gene*, pp. 105–06.

165 ". . . elephant seal . . ." *Ibid.*, p. 74.

165 Haldane quoted in Dawkins, *The Selfish Gene*, p. 103.

166 "Scientists who . . ." Fred Bruemmer, "White Whales on Holiday," *Natural History* (January 1986) p. 48.

166 "An Atlantic bottle-nosed . . ." Connor and Norris, "Are Dolphins Reciprocal Altruists?" p. 368.

166 "Washoe, the famous chimpanzee . . ." Told in Eugene Linden, *Silent Part-*

ners: The Legacy of the Ape Language Experiments (New York: Times Books, 1986), pp. 42–43. Also interview with Roger Fouts by Susan McCarthy, December 10, 1993.

166 "Asked whether he . . ." "Ripples of Controversy After a Chimp Drowns," *New York Times*, October 16, 1990. (The chimpanzee who drowned, referred to in the *Times* headline, is not the same animal as the one saved.)

167 "The three belugas . . ." Bruemmer, "White Whales on Holiday," pp. 40–49.

167 Connor and Norris, *Ibid.*, pp. 358–74.

168 "Two reporters . . ." Michael Hutchins and Kathy Sullivan, "Dolphin Delight," *Animal Kingdom* (July/August 1989), pp. 47–53.

169 Mike Tomkies, *Out of the Wild* (London: Jonathan Cape, 1985), p. 197.

169 Ryden, *Lily Pond*, p. 217.

169 Moss, *Elephant Memories*, p. 84.

169 Barry Holstun Lopez, *Of Wolves and Men* (New York: Charles Scribner's Sons, 1978), p. 198.

169 "A rabbit in . . ." Göran Högstedt, "Adaptation unto Death: Function of Fear Screams," *American Naturalist* 121 (1983), pp. 562–70.

170 ". . . when lions hunt . . ." Schaller, *Serengeti Lion,* p. 254.

170 "A handful of studies . . ." Hannah M. H. Wu, Warren G. Holmes, Steven R. Medina, and Gene P. Sackett, "Kin Preference in Infant *Macaca nemestrina*," *Nature* 285 (1980), pp. 225–27.

171 "However, it has . . ." Chadwick, *Beast the Color of Winter*, p. 15.

171 "Experimenters studying . . ." Robert M. Seyfarth and Dorothy L. Cheney, "Grooming, Alliances, and Reciprocal Altruism in Vervet Monkeys," *Nature* 308, No. 5 (April 1984), pp. 541–42.

171 ". . . to monitor their indebtedness . . ." Eugene S. Morton and Jake Page, *Animal Talk: Science and the Voices of Nature* (New York: Random House, 1992), pp. 138–39.

172 Joseph Wood Krutch, *The Best of Two Worlds* (New York: William Sloane Associates, 1950), p. 77.

172 "In the Negev Desert . . ." Clark, *High Hills and Wild Goats*, p. 136.

173 Athan, *Guide to a Well-Behaved Parrot*, pp. 111–12. Also interview by Susan McCarthy, August 23, 1993.

173 "In the Kenya bush . . ." Rasa, *Mongoose Watch*, pp. 83–84.

173 Thomas, *The Tribe of Tiger*, p. 25.

174 "Ola, a young false killer . . ." Pryor, *Lads Before the Wind*, pp. 218–19.

175 " '. . . doing her own genes no good . . .' " Dawkins, *Selfish Gene*, p. 109.

175 " '. . . that has no place in nature . . .' " *Ibid.*, p. 215.

175 "A recent scientific report . . ." Gerald S. Wilkinson, "Food Sharing in Vampire Bats," *Scientific American* 262 (1990), pp. 76–82. Also Gerald Wilkinson, interview by Susan McCarthy, March 4, 1994.

176 "Robert Frank . . ." Quoted in Kohn, *The Brighter Side of Human Nature*, p. 188.

176 Connor and Norris, "Are Dolphins Reciprocal Altruists?" pp. 358–74.

176 Robert L. Trivers, "The Evolution of Reciprocal Altruism," *Quarterly Review of Biology* 46 (1971), pp. 35–57.

177 "They were martyrs." Jim Nollman, *Animal Dreaming: The Art and Science of Interspecies Communication* (Toronto and New York: Bantam Books, 1987), p. 59.

Chapter 9: Shame, Blushing, and Hidden Secrets

179 Jane Goodall, interview by Susan McCarthy, May 7, 1994.

180 "Shame. . . . most vividly remembered . . ." This observation was made by John McCarthy.

180 ". . . 'the master emotion' . . ." Robert Karen, "Shame," *Atlantic Monthly* 269 (February 1992), pp. 40–70.

180 Donald Nathanson, *Shame and Pride: Affect, Sex, and the Birth of the Self* (New York: W. W. Norton & Co., 1992), p. 142.

181 "If such apes are anesthetized . . ." Gordon Gallup, "Self-recognition in Primates: A Comparative Approach to the Bidirectional Properties of Consciousness," *American Psychologist* 32 (1977), pp. 329–38. Gallup tested paint on himself before applying it to chimps.

181 Kennedy, *New Anthropomorphism*, pp. 107–08.

181 "The chimpanzees Sherman and Austin . . ." Savage-Rumbaugh, *Ape Language*, pp. 308–14.

182 ". . . Yeroen was slightly injured . . ." De Waal, *Chimpanzee Politics*, pp. 47–48.

182 ". . . to save face." De Waal, *Peacemaking Among Primates*, pp. 238–39.

182 Craig Packer, "Male Dominance and Reproductive Activity in *Papio anubis*," *Animal Behavior* 27 (1979), pp. 37–45.

182 ". . . lion glanced around . . ." Schaller, *Serengeti Lion*, p. 268.

182 "A mountain goat who sees a predator . . ." Chadwick, *Beast the Color of Winter*, pp. 87–88.

182 "Koko['s] embarrassment." Patterson and Linden, *Education of Koko*, pp. 136–37.

183 ". . . Wela, was trained . . ." Pryor, *Lads Before the Wind*, p. 128.

184 ". . . Washoe were human." Roger Fouts, interview by Susan McCarthy, December 10, 1993.

184 "One text on animal behavior evades . . ." Erika K. Honoré and Peter H. Klopfer, *A Concise Survey of Animal Behavior* (San Diego, CA: Academic Press/Harcourt Brace Jovanovich, 1990), p. 85.

184 The pre-Darwinian naturalist Jean de Lamarck (1744–1829) theorized that animals could inherit acquired characteristics, in this case the habit of blushing.

185 Darwin, *Expression of the Emotions*, p. 309.

185 "His data contradicted those defenders of slavery . . ." Nathanson, *Shame and Pride: Affect, Sex, and the Birth of the Self*, p. 462.

185 "Monkeys redden . . ." Darwin, *Expression of the Emotions*, p. 344.

185 "The ears of a Tasmanian devil . . ." Grzimek, *Animal Life Encyclopedia*, Vol. 10, p. 82.

185 ". . . smoky honeyeater have . . ." Bruce M. Beehler, *A Naturalist in New Guinea* (Austin, TX: University of Texas Press, 1991), p. 57.

185 Athan, *Guide to a Well-Behaved Parrot*, p. 13. Also interview by Susan McCarthy, August 23, 1993.

186 ". . . toys . . . designed to fall apart . . ." Cited in Lewis, *Shame*, pp. 5–26.

186 Helen Block Lewis cited in Nathanson, *Shame and Pride*, p. 218.

186 ". . . flashing colored lights . . ." *Ibid.*, pp. 169–70.

187 ". . . 'a biological system . . .'" *Ibid.*, p. 140.

187 ". . . global self-accusation . . ." *Ibid.*, pp. 210–11.

187 ". . . bone marrow of wildebeest . . ." Schaller, *Serengeti Lion*, p. 231.

187 "One experimenter who was shooting . . ." Kruuk, *Spotted Hyena*, pp. 99–100, 150.

188 ". . . acting oddly . . ." *Ibid.*, pp. 153–55.

188 ". . . small silver fish . . ." Cited in Norris, *Dolphin Days*, p. 188.

188 Leyhausen, *Cat Behavior*, pp. 144–45.

189 "Scottish red deer . . ." Darling, *Herd of Red Deer*, p. 81.

190 ". . . elephants who are laughed at . . ." David Gucwa and James Ehmann, *To Whom It May Concern: An Investigation of the Art of Elephants* (New York: W. W. Norton & Co., 1985), p. 200.

190 Terrace, *Nim*, pp. 222–26.

191 ". . . dogs do feel remorse . . ." Desmond Morris, *Dogwatching* (London: Jonathan Cape, 1986), p. 29.

191 Nathanson, *Shame and Pride*, p. 15.

Chapter 10: Beauty, the Bears, and the Setting Sun

192 Geza Teleki, "They Are Us," pp. 296–302 in *The Great Ape Project*.

192 "The sense of . . ." Izard, however, considers creativity part of an Interest-Excitement emotional complex, along with hope. Izard, *Human Emotions*, p. 42.

192 Adriaan Kortlandt, "Chimpanzees in the Wild," *Scientific American* 206 (May 1962), pp. 128–38.

193 ". . . animals . . . are color-blind." Paul Dickson and Joseph C. Gould, *Myth-Informed: Legends, Credos, and Wrongheaded "Facts" We All Believe* (Perigee/Putnam, 1993), p. 21. The authors write "bulls, like many other animals, including dogs, see only shades of light and dark." Also see John Horgan, "See Spot See Blue: Curb that Dogma! Canines Are Not Color-blind," *Scientific American* 262 (January 1990), p. 20. Horgan notes this assertion making its way into textbooks.

194 ". . . see ultraviolet light . . ." Gerald Jacobs, interview by Susan McCarthy, September 30, 1993.

194 ". . . magnetic fields." Wolfgang Wiltschko, Ursula Munro, Hugh Ford, and Roswitha Wiltschko, "Red Light Disrupts Magnetic Orientation of Migratory Birds," *Nature* 364 (August 5, 1993), p. 525.

194 " 'What is it about . . .'" Benyus, *Beastly Behaviors*, p. 206.

195 K. von Frisch: "Ein Zwergwels, der kommt, wenn man ihm pfeift." *Biologisches Zentralblatt* 43 (1923), pp. 439–46. In this article, von Frisch left it open whether the fish "heard" or "felt" the whistling. But in a later article (*Nature*, 141, January 1, 1938, pp. 8–11) he proved that they did hear. He noted that people had long believed fish made no sounds. Yet minnows make soft piping noises: "It is interesting that the production of sound by so well known a fish should have been overlooked for so long. There may be much to discover in the future about the language of fishes."

195 "Birds are about . . ." Joel Carl Welty and Luis Baptista, *The Life of Birds* (New York: Saunders College Publishing, 1988), pp. 82, 215. Among the functions of song, the authors note, "that some birds may sing from a sense of well-being, or simply 'for the joy of it,' should not arbitrarily be ruled out!"

195 Gerald Durrell, *My Family and Other Animals* (New York: Viking Press, 1957), pp. 38–39.

196 "Grey parrots . . ." De Grahl, *Grey Parrot*, p. 168.

196 Ryden, *God's Dog*, p. 70.

196 "In most gibbon . . ." Donna Robbins Leighton, "Gibbons: Territoriality and Monogamy" in *Primate Societies*, Barbara B. Smuts, Dorothy L. Cheney, Robert M. Seyfarth, Richard W. Wrangham, and Thomas T. Struhsaker, eds. (Chicago and London: University of Chicago Press, 1986), pp. 135–45.

196 Nollman, *Animal Dreaming*, pp. 94–97.

197 "Michael, a gorilla . . ." Wendy Gordon (Gorilla Foundation, Woodside, CA), interview by Susan McCarthy, April 29, 1994.

197 "Siri, an Indian . . ." Gucwa and Ehmann, *To Whom It May Concern*, p. 190.

198 ". . . coatimundis in Arizona . . ." Gilbert, *Chulo*, p. 202.

198 ". . . vultures . . ." Welty and Baptista, *Life of Birds*, pp. 78–79.

198 ". . . rhesus monkeys . . ." N. K. Humphrey, " 'Interest' and 'Pleasure': Two Determinants of a Monkey's Visual Preferences," *Perception* 1 (1972), pp. 395–416.

199 Bernhard Rensch cited in Desmond Morris, *The Biology of Art: A Study of the Picture-Making Behavior of the Great Apes and Its Relationship to Human Art* (New York: Alfred A. Knopf, 1962), pp. 32–34.

199 ". . . widowbirds . . ." Malte Anderson, "Female Choice Selects for Extreme Tail Length in a Widowbird," *Nature* 299 (October 28, 1982), pp. 818–20.

199 ". . . bowerbirds and birds of paradise . . ." Both are members of the family Paradiseidae.

199 ". . . the satin bowerbird . . ." Welty and Baptista, *Life of Birds*, pp. 278–80.

200 Beehler, *Naturalist in New Guinea*, p. 45.

201 " '. . . control of the lek . . .' " *Ibid.*, p. 147.

201 "New Guinea human . . ." It is also worth keeping in mind that plumage may send a message to someone other than a potential mate or rival. Beehler and colleagues recently made the discovery that the hooded pitohui, also of New Guinea, has a powerful neurotoxin in its bright orange and black feathers, which is believed to protect it from predators. Here the plumage pre-

sumably has, at least in part, a warning message. See John P. Dumbacher, Bruce M. Beehler, Thomas F. Spande, H. Martin Garaffo, and John W. Daly, "Homobatrachotoxin in the Genus *Pitohui:* Chemical Defense in Birds?" *Science* 258 (October 30, 1992), pp. 799–801. Natives of New Guinea have long known that pitohuis have "bitter" skin.

202 "Alpha, a chimp . . ." Paul H. Schiller, "Figural Preferences in the Drawings of a Chimpanzee," *Journal of Comparative and Physiological Psychology* 44 (1951), pp. 101–11.

202 Desmond Morris, *Animal Days* (London: Jonathan Cape, 1979), pp. 197–98. Also Morris, *The Biology of Art: A Study of the Picture-Making Behavior of the Great Apes and Its Relationship to Human Art.*

202 "One chimp, Moja . . ." Kathleen Beach, Roger S. Fouts, and Deborah H. Fouts, "Representational Art in Chimpanzees," *Friends of Washoe* 3 (Summer 1984), pp. 2–4; Roger Fouts interview; also A. Gardner and B. Gardner, "Comparative Psychology and Language Acquisition," *Annals of the New York Academy of Sciences* 309 (1978), pp. 37–76. Cited in Gucwa and Ehmann.

205 Gucwa and Ehmann, *To Whom It May Concern*, pp. 119–20.

206 ". . . San Diego Zoo . . ." *Ibid.*, pp. 93–97.

206 Chadwick, *The Fate of the Elephant*, pp. 12–15.

207 Pryor, *Lads Before the Wind*, pp. 234–53; Karen Pryor, Richard Haag, and Joseph O'Reilly, "The Creative Porpoise: Training for Novel Behavior," *Journal of the Experimental Analysis of Behavior* 12 (1969), pp. 653–61. For the journal article all references to Hou as "she" were changed to "it."

It is interesting to see how anecdotal information was transformed into acceptable data in this story. Despite the presence of careful observers, Malia's creativity had the status of an anecdote. Hou's almost identical display of creativity was not, presumably in large part *because it was expected.* The recording of Hou's actions on film is irrelevant: the vast majority of animal behavior that makes its way into the literature is not documented in this way.

209 "In Japanese monkey troops . . ." Toshisada Nishida, "Local Traditions and Cultural Transmission," in *Primate Societies*, Barbara B. Smuts, Dorothy L. Cheney, Robert M. Seyfarth, Richard W. Wrangham and Thomas T. Struhsaker, eds. (Chicago and London: University of Chicago Press, 1986), pp. 462–74; Marvin Harris, *Our Kind* (New York: Harper & Row, 1989), p. 63.

209 Thomas, "Reflections: The Old Way."

209 ". . . olive baboons . . ." Strum, *Almost Human*, pp. 128–33. This tradition of intensive hunting for meat later vanished.

209 "A curious example . . ." De Waal, *Chimpanzee Politics*, p. 135.

210 "In the University of Washington . . ." Roger S. Fouts and Deborah H. Fouts, "Chimpanzees' Use of Sign Language," 28–41 in Cavalieri and Singer, eds., *The Great Ape Project*, pp. 37–38.

210 ". . . 'just interesting oddities.' " Dawkins, *Selfish Gene*, pp. 203–04.

210 "Cultural transmission may . . ." Richard Dawkins coined the word *meme* to mean a bit or collection of bits of information that are behaviorally transferred from one individual to another, including such things as tunes, techniques, fashions, and phrases. It is part of the current scientific fashion to

ascribe a great deal of human and animal behavior to genetic causes. Until we are better able to tell a meme from a gene, such conclusions are often unwarranted.

211 Krutch, *Best of Two Worlds*, pp. 92–94.

211 ". . . elephants in Kenya . . ." Chadwick, *The Fate of the Elephant*, p. 63.

Chapter 11: The Religious Impulse, Justice, and the Inexpressible

212 "We are all . . ." Darwin once wrote himself a note: "Never use the words *higher* and *lower*." *More Letters of Charles Darwin*, edited by F. Darwin and A. C. Seward (London: Murray, 1903), Vol. 1, p. 114n.

213 Jolly, *Lemur Behavior*, p. 36.

213 ". . . awe as a form of shame." Nathanson, *Shame and Pride*, p. 474.

213 Thomas, *Hidden Life of Dogs*, xvii–xviii.

214 "Nim Chimpsky learned . . ." Terrace, *Nim*, p. 171.

214 De Waal, *Chimpanzee Politics*, pp. 171–72.

214 "In another incident . . ." *Ibid.*, p. 207.

215 Thomas, *Hidden Life of Dogs*, pp. 49–51.

215 ". . . coatimundis . . ." Gilbert, *Chulo*, p. 105–06.

216 Terrace, *Nim*, pp. 185–86.

216 "Signing apes have . . ." The chimpanzees in the later sign language projects of the Gardners and the gorillas taught by Patterson did have native signers among their teachers. In no case were the lead researchers fluent signers, however.

217 Terrace, *Nim*, Appendix B, "Recruiting Nim's Teachers," pp. 392–95.

217 "Moja, who knows . . ." Roger Fouts, interview by Susan McCarthy, December 10, 1993.

217 Donald R. Griffin: "The Cognitive Dimensions of Animal Communication," in *Fortschritte Der Zoologie*, 31 (1985), pp. 471–82.

218 Savage-Rumbaugh, *Ape Language*, p. 337.

218 Jim Nollman, *Animal Dreaming: The Art and Science of Interspecies Communication*, p. 105. Cf. the authoritative article in *The Encyclopedia of Mammals*: "It is clear from its continuous nature and ordered sequence that the song potentially contains much information, but its precise function is not known. Most evidence at present indicates that the prime function of the song is sexual." David Macdonald, ed. (New York: Facts on File Publications, 1984), p. 229.

219 Schaller, *Serengeti Lion*, p. 50.

220 Joyce Poole quoted in Chadwick, *Fate of the Elephant*, pp. 75–76.

220 " 'On meeting a gorilla . . .' " George B. Schaller, *The Last Panda* (Chicago and London: University of Chicago Press, 1972), pp. 79–80.

221 " 'Certain usages . . .' " Terrace, *Nim*, pp. 222–26.

222 "As Julian Huxley . . ." Quoted in Krutch, *Best of Two Worlds*.

223 Joseph Wood Krutch, *The Great Chain of Life* (Boston: Houghton Mifflin, 1956), p. 106.

224 Lynn Rogers, interviews by Susan McCarthy, July 15, 1993, and May 11, 1994.

225 Thomas, "Reflections: The Old Way," p. 100.

Conclusion: Sharing the World with Feeling Creatures

227 Rousseau from Lester G. Crocker, ed., *The Social Contract and Discourse on the Origin and Foundation of Inequality Among Mankind* (New York: Washington Square Press, 1967), p. 172.

227 Brigid Brophy: "In Pursuit of a Fantasy," in *Animals, Men and Morals*, pp. 125–45, S. and R. Godlovitch, eds. (New York: Taplinger Publishing Co., 1972), p. 129.

227 The quote is from Darwin's *The Descent of Man*. Quoted here from Marian Scholtmeijer, *Animal Victims in Modern Fiction: From Sanctity to Sacrifice* (Toronto: University of Toronto Press, 1993).

228 First published in the *American Scholar*, Vol. 40, No. 3 (Summer 1971) as "Antivivisection: The Reluctant Hydra," and reprinted with the title "A Defense of Vivisection" in *Animal Rights and Human Obligations*, Tom Regan and Peter Singer, eds. (New Jersey: Prentice-Hall, 1976), pp. 163–69.

228 "As late as the 1930s . . ." No doubt he did. He had even greater problems. At the end of his book *Man the Unknown* (New York & London: Harper, 1935), on p. 318, this Nobel laureate wrote:
"There remains the unsolved problem of the immense number of defectives and criminals. They are an enormous burden for the part of the population that has remained normal. Gigantic sums are now required to maintain prisons and insane asylums and protect the public against gangsters and lunatics. Why do we preserve these useless and harmful beings? The abnormal prevent the development of the normal. Why should society not dispose of the criminals and the insane in a more economical manner? . . . Perhaps prisons should be abolished. They could be replaced by smaller and less expensive institutions. The conditioning of petty criminals with the whip, or some more scientific procedure, followed by a short stay in hospital, would probably suffice to insure order. Those who have murdered, robbed while armed with automatic pistol or machine gun, kidnapped children, despoiled the poor of their savings, misled the public in important matters, should be humanely and economically disposed of in small euthanasic institutions supplied with proper gases. A similar treatment could be advantageously applied to the insane, guilty of criminal acts. Modern society should not hesitate to organize itself with reference to the normal individual. Philosophical systems and sentimental prejudices must give way before such a necessity." Hitler's doctor, Karl Brandt, in his trial in Nuremberg, offered this book in his defense.

229 "When animal psychologist . . ." S. Begley and J. Cooper Ramo, "Not Just a Pretty Face," *Newsweek* (November 1, 1993), p. 67.

229 "Recently a steer . . ." "Steer Flees Slaughter and Is Last Seen Going Thataway," *New York Times*, May 24, 1990.

230 "Perhaps just . . ." A German woman leaving a theater performance of *The Diary of Anne Frank* was heard to say to her companion: "That one, at least, should not have been killed."

230 "Modern philosophers . . ." The new school of cognitive ethology, started by Donald R. Griffin, is an exception, and many of the biologists and animal behaviorists who work in this area, people such as Gordon Burghardt, Dorothy Cheney, Robert Seyfarth, Carolyn Ristau, Marc Bekoff, Dale Jamieson, Alison Jolly and many others, would agree with the position that animals lead emotional lives, though they might not all agree on how complex and sophisticated they are.

230 This passage from *Introduction to the Principles of Morals and Legislation* by Jeremy Bentham (Chapter 18, sec. 1) as well as selections from his "A Utilitarian View" and John Stuart Mill's "A Defence of Bentham," can be found in the useful collection edited by P. Singer and T. Regan, *Animal Rights and Human Obligations* (Englewood Cliffs, NJ: Prentice-Hall, 1976).

231 ". . . to a spider . . ." P. N. Witt, "Die Wirkung einer einmaligen Gabe von Largactil auf den Netzbau Der Spinne Zilla-x-notata," in *Monatschrift fuer Psychiatrie und Neurologie*, 129 (1955), Nos. 1–3, pp. 123–28.

232 Roberts, *The Scientific Conscience: Reflections on the Modern Biologist and Humanism*, pp. 106–07.

232 " 'The details of his . . .' " Quoted by Dr. White himself on p. 166 of the article cited earlier in this chapter.

232 Goodall in *The Great Ape Project*, Cavalieri and Singer, eds., pp. 15–16.

233 Chadwick, *The Fate of the Elephant*. Quoted by E. M. Thomas in "The Battle for the Elephants," *New York Review of Books* (March 24, 1994), p. 5.

233 " 'In the distance . . .' " Somadeva, *Kathasaritsagara*, Durgaprasad Parab, ed. (Bombay, India: Nirnaya Sagara Press, 1903), Ch. 64, vv. 4–12. See, too, the 12-volume translation of *The Ocean of Story*, translated by C. H. Tawney, edited by N. M. Penzer (London: Chas. J. Sawyer), Vol. 5, 1926, pp. 138ff. The editor, p. 34 of the Introduction, notes that "India is indeed the home of storytelling. It was from here that the Persians learned the art, and passed it on to the Arabians. From the Middle East the tales found their way to Constantinople and Venice, and finally appeared in the pages of Boccaccio, Chaucer and La Fontaine." The story is old, probably predating the Christian era, being found in the Sanskrit Pancatantra. (In the Pancatantra version, there are a few more details about the mongoose: "He left behind a mongoose that he had raised just like a son, keeping him in his house in the room where the sacred fire was kept and feeding him on kernels of corn and the like." (Franklin Edgerton: *The Pancatantra Reconstructed*. Vol. 2: Introduction and Translation. New Haven: American Oriental Society, 1924. p. 403.)

234 Jan Harold Brunvand, *The Mexican Pet* (New York: W. W. Norton, 1986), p. 44.

234 "We cannot know . . ." See M. B. Emeneau, "A Classical Indian Folk-Tale as a Reported Modern Event: The Brahman and the Mongoose." *Proceedings of the American Philosophical Society*, Vol. 83, No. 3, September 1940, pp. 503–13. This reports a modern event that mirrors the classic story. Emeneau

concludes that the modern report "is one of actual events." On August 17, 1994, I spoke with Professor Emeneau (now in his nineties). He told me that his associate, the anthropologist David Mandelbaum, interviewed the woman whose mongoose it was. Emeneau was working with these hill tribes, the Kotas of the Nilgiris in South India during 1935–38. He told me that because the storytellers incorporate so much material from the plains (the Kotas live on a 7,000-foot plateau), including literary motifs, it is impossible to know for certain whether the event actually happened or not. He has changed his mind several times over the years, and is now unable to decide whether it did or did not happen. The woman, however, claims to have been an eyewitness—in fact, the protagonist of the story (in the modern version she kills the mongoose). Legend or fact, the story resonates with the modern reader, at least with this modern reader.

234 *The Attic Nights of Aulus Gellius,* with an English translation by John C. Rolfe. 3 vols. (Cambridge, MA: Harvard University Press, 1984, vol. 1, pp. 421–27). Only fragments of the *Wonders of Egypt* exist. A very similar account, from the same source, was made famous in Europe in the sixteenth century by Michel Montaigne. See *The Complete Essays of Montaigne,* trans. by Donald M. Frame (Stanford, CA: Stanford University Press, 1989, pp. 350–51). For more on Apion, see Pauly's *Realencyclopädie der classischen Altertumswissenschaft,* Vol. 1, pt. 2 (1894), Article on Apion 3, esp. p. 2805. The very fact that he claimed to have been an eyewitness to the event is used, here, against his credibility.

236 "'This is the lion . . .'" There is considerable literature on this topic. See August Marx, *Griechische Märchen von dankbaren Tieren und verwandtes* (Stuttgart: Verlag von W. Kohlhammer, 1889). On p. 58 he points out that the famous Brehma (Tierl. I, pp. 369 and 378) leaves it unclear whether he thinks the story of Androcles is possible or not. St. Hieronymus also takes a thorn out of the paw of a lion (p. 61 for sources). See, too, the excellent book by Otto Keller: *Thiere des classischen Alterthums in culturgeschichtlicher Beziehung* (Innsbruck: Verlag der Wagner'schen Universitätsbuchhandlung, 1887). He is especially good on dolphins.

236 "Is this fiction . . ." See Adrian House, *The Great Safari: The Lives of George and Joy Adamson, Famous for Born Free* (New York: William Morrow & Co., 1993). In her original book, *Born Free: A Lioness of Two Worlds* (Fontana/ Collins Harvill, 1960, p. 49), Joy Adamson notes: "Most lions take to man-eating because they have some infirmity: either they have been wounded by an arrow head or damaged in a trap, or their teeth are in bad condition, or they have porcupine quills in their paws."

236 "An Interview with John Lilly," *New Frontier,* September 1987, p. 10.

Bibliography

Adams, Jack. *Wild Elephants in Captivity.* Dominguez Hills, CA: Center for the Study of Elephants, 1981.

Alcock, John. *Animal Behavior: An Evolutionary Approach,* 4th ed. Sunderland, MA: Sinauer Associates, 1989.

Alpers, Antony. *Dolphins: The Myth and the Mammal.* Boston: Houghton Mifflin Co., 1960.

Anand, K. J. S., and McGrath, P. J., eds. *Pain in Neonates.* Amsterdam: Elsevier, 1993.

Anderson, Malte. "Female Choice Selects for Extreme Tail Length in a Widowbird." *Nature* 299 (1982): 818–20.

Angier, Natalie. "Dolphin Courtship: Brutal, Cunning and Complex." *New York Times,* February 18, 1992.

Archibald, George. "Gee Whiz! ICF Hatches a Whooper." *The ICF Bugle* (July 1982): 1.

Arzt, Volker, and Birmelin, Immanuel. *Haben Tieren ein Bewusstsein?: Wenn Affen lügen, wenn Katzen denken und Elefanten traurig sind.* Munich: C. Bertelsmann, 1993.

Athan, Mattie Sue. *Guide to a Well-Behaved Parrot.* Hauppauge, NY: Barron's Educational Series, 1993.

Bachmann, Christian, and Kummer, Hans. "Male Assessment of Female Choice in Hamadryas Baboons." *Behavioral Ecology and Sociobiology* 6 (1980): 315–21.

Bailey, Robert O.; Seymour, Norman R.; and Stewart, Gary R. "Rape Behavior in Blue-winged Teal." *Auk* 95 (1978): 188–90.

Baptista, Luis. "A Letter from the Field." *Pacific Discovery* 16 (4): 44–47.

Barash, David P. "Sociobiology of Rape in Mallards *(Anas platyrhynchos):* Responses of the Mated Male." *Science* 197 (August 19, 1977): 788–89.

Barber, Theodore Xenophon. *The Human Nature of Birds: A Scientific Discovery with Startling Implications.* New York: St. Martin's Press, 1993.

Barinaga, Marcia. "How Scary Things Get That Way." *Science* 258 (November 6, 1992): 887–88.

Beach, Kathleen; Fouts, Roger S.; and Fouts, Deborah H. "Representational Art in Chimpanzees." *Friends of Washoe* 3 (Summer 1984): 2–4.

Beehler, Bruce M. *A Naturalist in New Guinea.* Austin, TX: University of Texas Press, 1991.

Begley, Sharon, and Ramo, Joshua Cooper. "Not Just a Pretty Face." *Newsweek* (November 1, 1993): 67.

Benyus, Janine M. *Beastly Behaviors: A Zoo Lover's Companion: What Makes Whales Whistle, Cranes Dance, Pandas Turn Somersaults, and Crocodiles Roar: a Watcher's Guide to How Animals Act and Why.* Reading, MA: Addison-Wesley Publishing Co., 1992.

Berggren, Sigvard. *Berggren's Beasts.* Translated by Ian Rodger. New York: Paul S. Eriksson, 1970.

Bernstein, Irwin S. "Dominance: The Baby and the Bathwater." *Behavioral and Brain Sciences* 4 (1981): 419–29.

Bledsoe, Thomas. *Brown Bear Summer: My Life Among Alaska's Grizzlies.* New York: Dutton, 1987.

Boas, George. "The Happy Beast." In *French Thought of the Seventeenth Century: Contributions to the History of Primitivism.* Baltimore: The Johns Hopkins Press, 1933.

Boswall, Jeffery. "Russia Is for the Birds." *Discover* (March 1987): 78–83.

Broun, Heywood Hale. "Ever Indomitable, Secretariat Thunders Across the Ages." *New York Times,* May 30, 1993.

Bruemmer, Fred. "White Whales on Holiday." *Natural History* (January 1986): 41–49.

Bullard, Edward. "The Emergence of Plate Tectonics: A Personal View." *Annual Review of Earth and Planetary Science* 3 (1975): 1–30.

Burghardt, Gordon M. "Animal Awareness: Current Perceptions and Historical Perspective." *American Psychologist* 40 (August 1985): 905–19.

Buyukmihci, Hope Sawyer. *The Hour of the Beaver.* Chicago: Rand McNally & Co., 1971.

Callwood, June. *Emotions: What They Are and How They Affect Us, from the Basic Hates and Fears of Childhood to More Sophisticated Feelings That Later Govern Our Adult Lives: How We Can Deal with the Way We Feel.* Garden City, NY: Doubleday & Co., 1986.

Campbell, Robert Jean. *Psychiatric Dictionary,* 5th ed. New York and Oxford: Oxford University Press, 1981.

Candland, Douglas Keith. *Feral Children & Clever Animals: Reflections on Human Nature.* Oxford, England: Oxford University Press, 1993.

Carlquist, Sherwin. *Island Life: A Natural History of the Islands of the World.* Garden City, NY: Natural History Press, 1965.

Carrington, Richard. *Elephants: A Short Account of Their Natural History, Evolution and Influence on Mankind.* London: Chatto and Windus, 1958.

Carson, Gerald. *Men, Beasts, and Gods: A History of Cruelty and Kindness to Animals.* New York: Charles Scribner's Sons, 1972.

Cartmill, Matt. *A View to a Death in the Morning: Hunting and Nature Through History.* Cambridge, MA: Harvard University Press, 1993.

Cavalieri, Paola, and Singer, Peter, eds. *The Great Ape Project: Equality Beyond Humanity.* London: Fourth Estate, 1993.

Chadwick, Douglas H. *A Beast the Color of Winter: The Mountain Goat Observed.* San Francisco: Sierra Club Books, 1983.

———. *The Fate of the Elephant.* San Francisco: Sierra Club Books, 1992.

Chalmers, N. R. "Dominance as Part of a Relationship." *Behavioral and Brain Sciences* 4 (1981): 437–38.

Clark, Bill. *High Hills and Wild Goats.* Boston: Little, Brown and Co., 1990.

Cochrane, Robert. "Working Elephants at Rangoon," "Some Parrots I Have Known." In *The Animal Story Book,* Vol. IX of The Young Folks Library. Boston: Hall & Locke Co., 1901.

Colmenares, F., and Rivero, H. "Male-Male Tolerance, Mate Sharing and Social Bonds Among Adult Male Brown Bears Living under Group Conditions in Captivity." *Acta Zoologica Fennica* 174 (1983): 149–51.

Connor, Richard C., and Norris, Kenneth S. "Are Dolphins Reciprocal Altruists?" *The American Naturalist* 119, #3 (March 1982): 358–74.

Conover, Adele. "He's Just One of the Bears." *National Wildlife* (June–July 1992): 30–36.

Crisler, Lois. *Captive Wild.* New York: Harper & Row, 1968.

Crumley, Jim. *Waters of the Wild Swan.* London: Jonathan Cape, 1992.

Dagg, Anne Innis, and Foster, J. Bristol. *The Giraffe: Its Biology, Behavior, and Ecology.* New York: Van Nostrand Reinhold Co., 1976.

Dagognet, François. *Traité des animaux.* Paris: Librairie Philosophique J. Vrin, 1987.

Darling, F. Fraser. *A Herd of Red Deer: A Study in Animal Behavior.* London: Oxford University Press, 1937.

Darwin, Charles. *The Descent of Man; and Selection in Relation to Sex.* 1871; reprint, with preface by Ashley Montagu, Norwalk, CT: Heritage Press, 1972.

———. *The Expression of the Emotions in Man and Animals.* 1872; reprint, Chicago and London: University of Chicago Press, 1965.

———. *The Correspondence of Charles Darwin. Volume 2; 1837–1843.* Cambridge, England: Cambridge University Press, 1986.

Dawkins, Richard. *The Selfish Gene.* New York: Oxford University Press, 1976.

De Grahl, Wolfgang. *The Grey Parrot.* Translated by William Charlton. Neptune City, NY: T.F.H. Publications, 1987.

De Rivera, Joseph. *A Structural Theory of the Emotions.* New York: International Universities Press, 1977.

De Waal, Frans. *Chimpanzee Politics: Power and Sex Among Apes.* New York: Harper & Row, 1982.

———. *Peacemaking Among Primates.* Cambridge, MA and London: Harvard University Press, 1989.

Dews, Peter B. "Some Observations on an Operant in the Octopus." *Journal of the Experimental Analysis of Behavior* 2 (1959): 57–63. 1959. *Readings in Animal Behavior*, edited by Thomas E. McGill. New York: Holt, Rinehart and Winston, 1965.

Dickson, Paul, and Gould, Joseph C. *Myth-Informed: Legends, Credos, and Wrong-headed "Facts" We All Believe*. Perigee/Putnam, 1993.

Dumbacher, John P.; Beehler, Bruce M.; Spande, Thomas F.; Garaffo, H. Martin; and Daly, John W. "Homobatrachotoxin in the Genus *Pitohui:* Chemical Defense in Birds?" *Science* 258 (October 30, 1992): 799–801.

Durrell, Gerald. *Menagerie Manor*. New York: Avon, 1964.

———. *My Family and Other Animals*. New York: Viking Press, 1957.

Emde, R. N., and Koenig, K. L. "Neonatal Smiling and Rapid Eye-movement States." *Journal of the American Academy of Child Psychiatry* 8 (1969): 57–67.

Emeneau, Murray B. "A Classical Indian Folk-Tale as a Reported Modern Event: The Brahman and the Mongoose." *Proceedings of the American Philosophical Society* 83, No. 3 (September 1940), 503–13.

Emlen, S. T., and Wrege, P. H. "Forced Copulations and Intraspecific Parasitism: Two Costs of Social Living in the White-fronted Bee-eater." *Ethology* 71 (1986): 2–29.

Fadiman, Anne. "Musk Ox Ruminations." *Life* (May 1986): 95–110.

Fagen, Robert. *Animal Play Behavior*. New York and Oxford: Oxford University Press, 1981.

Fisher, John Andrew. "Disambiguating Anthropomorphism: An Interdisciplinary Review." In *Perspectives in Ethology* 9 (1991).

Fossey, Dian. *Gorillas in the Mist*. Boston: Houghton Mifflin Co., 1983.

Fouts, Roger S.; Fouts, Deborah H.; and Van Cantfort, Thomas E. "The Infant Loulis Learns Signs from Cross-Fostered Chimpanzees." In *Teaching Sign Language to Chimpanzees*, edited by R. Allen Gardner, Beatrix T. Gardner, and Thomas E. Van Cantfort. Albany: State University of New York Press, 1989.

Fowles, John, with Horvat, Frank. *The Tree*. Boston: Little, Brown & Co., 1979.

Fox, Michael W. *The Whistling Hunters: Field Studies of the Asiatic Wild Dog (Cuon alpinus)*. Albany: State University of New York Press, 1984.

Frank, Robert. *Passions Within Reason: The Strategic Role of the Emotions*. New York: Norton, 1988.

French, Thomas, M. *The Integration of Behavior, Volume 1: Basic Postulates*. Chicago: University of Chicago Press, 1952.

Frey, William H., II, with Langseth, Muriel. *Crying: The Mystery of Tears*. Minneapolis: Harper & Row/Winston Press, 1985.

Gallup, Gordon. "Self-recognition in Primates: A Comparative Approach to the Bidirectional Properties of Consciousness." *American Psychologist* 32 (1977): 329–38.

Gallup, Gordon G., and Suarez, Susan D. "Overcoming Our Resistance to Animal Research: Man in Comparative Perspective." In *Comparing Behavior: Studying Man Studying Animals*, edited by D. W. Rajecki. Hillsdale, NJ: Lawrence Erlbaum Associates, 1983.

Gardner, R. Allen, and Gardner, Beatrix T. "Comparative Psychology and Language Acquisition." *Annals of the New York Academy of Sciences* 309 (1978): 37–76.

———. "A Cross-Fostering Laboratory." In *Teaching Sign Language to Chimpanzees.* See Fouts, 1989.

Gardner, Beatrix T.; Gardner, R. Allen; and Nichols, Susan G. "The Shapes and Uses of Signs in a Cross-Fostering Laboratory." In *Teaching Sign Language to Chimpanzees.* See Fouts, 1989.

Gauntlett, Ian S.; Koh, T. H. H. G.; and Silverman, William A. "Analgesia and Anaesthesia in Newborn Babies and Infants." (Letters) *Lancet,* May 9, 1987.

Gebel-Williams, Gunther, with Reinhold, Toni. *Untamed: The Autobiography of the Circus's Greatest Animal Trainer.* New York: William Morrow & Co., 1991.

Gilbert, Bil. *Chulo.* New York: Alfred A. Knopf, 1973.

Godlovitch, Stanley, and Godlovitch, Rosalind, eds. *Animals, Men and Morals.* New York: Taplinger Publishing Co., 1972.

Goodall, Jane. *The Chimpanzees of Gombe; Patterns of Behavior.* Cambridge, MA and London: Harvard University Press, 1986.

———. *In the Shadow of Man,* revised edition. Boston: Houghton Mifflin Co., 1988.

———. *Through a Window: My Thirty Years with the Chimpanzees of Gombe.* Boston: Houghton Mifflin Co., 1990.

———. *With Love.* Ridgefield, CT: Jane Goodall Institute, 1994.

Gormley, Gerard. *Orcas of the Gulf; a Natural History.* San Francisco: Sierra Club Books, 1990.

Gould, Stephen Jay. *The Mismeasure of Man.* New York: W. W. Norton & Co., 1981.

Griffin, Donald. *The Question of Animal Awareness: Evolutionary Continuity of Mental Experience.* New York: Rockefeller University Press, 1976. (2nd ed., 1981).

———. "Prospects for a Cognitive Ethology." *Behavioral and Brain Sciences* 1 (1978): 527–38.

———. *Animal Thinking.* Cambridge, MA: Harvard University Press, 1984.

———. "Animal Consciousness." *Neuroscience & Biobehavioral Reviews* 9 (1985): 615–22.

———. "The Cognitive Dimensions of Animal Communication." *Fortschritte der Zoologie* 31 (1985): 471–82.

———. *Animal Minds.* Chicago and London: University of Chicago Press, 1992.

Griffin, Donald, ed. *Animal Mind–Human Mind: Report of the Dahlem Workshop on Animal Mind–Human Mind, Berlin, 1981, March 22–27.* Berlin, NY: Springer-Verlag, 1982.

Grzimek, Bernhard, ed. *Grzimek's Animal Life Encyclopedia.* New York: Van Nostrand Reinhold Co., 1972.

Gucwa, David, and Ehmann, James. *To Whom It May Concern: An Investigation of the Art of Elephants.* New York: W. W. Norton & Co., 1985.

Gwinner, Eberhard, and Kneutgen, Johannes. "Über die biologische Bedeutung der 'zweckdienlichen' Anwendung erlernter Laute bei Vögeln." *Zeitschrift für Tierpsychologie* 19 (1962): 692–96.

Hall, Nancy. "The Painful Truth." *Parenting* (June/July 1992): 71–75.

Hargrove, Eugene C., ed. *The Animal Rights/Environmental Ethics Debate*. Albany: State University of New York Press, 1992.

Harlow, Harry. "The Nature of Love." *American Psychologist* 13 (1958): 673–85.

———. "Love in Infant Monkeys." *Scientific American* 200, #6 (1959): 68–74.

Harlow, Harry, and Suomi, Stephen J. "Depressive Behavior in Young Monkeys Subjected to Vertical Chamber Confinement." *Journal of Comparative and Physiological Psychology* 80 (1972): 11–18.

Harre, R., and Reynolds, V., eds. *The Meaning of Primate Signals*. Cambridge, England: Cambridge University Press, 1984.

Harris, Marvin. *Our Kind*. New York: Harper & Row, 1989.

Hastings, Hester. *Man and Beast in French Thought of the Eighteenth Century*, Vol. 27. Baltimore: The Johns Hopkins Press, 1936.

Hearne, Vicki. *Adam's Task: Calling Animals by Name*. New York: Alfred A. Knopf, 1986.

———. *Animal Happiness*. New York: HarperCollins, 1994.

Helfer, Ralph. *The Beauty of the Beasts: Tales of Hollywood's Animal Stars*. Los Angeles: Jeremy P. Tarcher, 1990.

Henderson, J. Y., with Taplinger, Richard. *Circus Doctor*. Boston: Little, Brown & Co., 1951.

Hill, Craven. "Playtime at the Zoo." *Zoo-Life* 1: 24–26.

Hinde, Robert A. *Animal Behavior*. New York: McGraw-Hill, 1966.

Hinsie, Leland E., and Campbell, Robert J. *Psychiatric Dictionary*, 4th ed. New York: Oxford University Press, 1970.

Högstedt, Göran. "Adaptation unto Death: Function of Fear Screams." *The American Naturalist* 121 (1983): 562–70.

Holt, Patricia. "Puppy Love Isn't Just for People: Author Says Dogs, Like Humans, Can Bond." *San Francisco Chronicle*, December 9, 1993.

Honoré, Erika K., and Klopfer, Peter H. *A Concise Survey of Animal Behavior*. San Diego, CA: Academic Press/Harcourt Brace Jovanovich, 1990.

Horgan, John. "See Spot See Blue: Curb That Dogma! Canines Are Not Colorblind." *Scientific American* 262 (January 1990): 20.

Hornocker, Maurice G. "Winter Territoriality in Mountain Lions." *Journal of Wildlife Management* 33 (July 1969): 457–64.

Hornstein, Harvey A. *Cruelty and Kindness: A New Look at Oppression and Altruism*. Englewood Cliffs, NJ: Prentice-Hall, 1976.

Houle, Marcy Cottrell. *Wings for My Flight: The Peregrine Falcons of Chimney Rock*. Reading, MA: Addison-Wesley Publishing Co., 1991.

Humphrey, N. K. " 'Interest' and 'Pleasure': Two Determinants of a Monkey's Visual Preferences." *Perception* 1 (1972): 395–416.

———. "The Social Function of Intellect." In *Growing Points in Ethology*, edited by P. P. G. Bateson and R. A. Hinde. Cambridge, England: Cambridge University Press, 1976: 303–17.

———. "Nature's Psychologists." In *Consciousness and the Physical World*, edited by B. D. Josephson and V. S. Ramachandran. Oxford, England: Pergamon Press, 1980: 57–80.

Hutchins, Michael, and Sullivan, Kathy. "Dolphin Delight." *Animal Kingdom* (July/August 1989).

Huxley, Thomas H. *Method and Results: Essays.* 1893; reprint, London: Macmillan, 1901.

Izard, Carroll E. *Human Emotions.* New York and London: Plenum Press, 1977.

Izard, Carroll E., and Buechler, S. "Aspects of Consciousness and Personality in Terms of Differential Emotions Theory." In *Emotion: Theory, Research, and Experience, Vol. I: Theories of Emotion,* edited by Robert Plutchik and Henry Kellerman. New York: Academic Press, 1980: 165–87.

Johnson, Dirk. "Now the Marlboro Man Loses His Spurs." *New York Times* (October 11, 1993): A1, A8.

Jolly, Alison. *Lemur Behavior: A Madagascar Field Study.* Chicago: University of Chicago Press, 1966.

Jordan, William. *Divorce Among the Gulls: An Uncommon Look at Human Nature.* San Francisco: North Point Press, 1991.

Josephson, B. D., and Ramachandran, V. S., eds. *Consciousness and the Physical World.* Oxford, England: Pergamon Press, 1980.

Karen, Robert. "Shame." *Atlantic Monthly* 269 (February 1992): 40–70.

Kavanau, J. Lee. "Behavior of Captive White-footed Mice." *Science* 155 (March 31, 1967): 1623–39.

Kellert, Stephen R., and Berry, Joyce K. *Phase III: Knowledge, Affection and Basic Attitudes Toward Animals in American Society.* U.S. Fish and Wildlife Service, 1980.

Kennedy, John S. *The New Anthropomorphism.* Cambridge, England: Cambridge University Press, 1992.

Kevles, Bettyann. *Females of the Species: Sex and Survival in the Animal Kingdom.* Cambridge, MA: Harvard University Press, 1986.

Kitcher, Philip. *Vaulting Ambition: Sociobiology and the Quest for Human Nature.* Cambridge, MA: MIT Press, 1985.

Kleiman, Devra G., and Malcolm, James R. "The Evolution of Male Parental Investment in Mammals." In *Parental Care in Mammals,* edited by David J. Gubernick and Peter H. Klopfer. New York: Plenum Press, 1981.

Kohn, Alfie. *The Brighter Side of Human Nature: Altruism and Empathy in Everyday Life.* New York: Basic Books, 1990.

Konner, Melvin. *The Tangled Wing: Biological Constraints on the Human Spirit.* New York: Holt, Rinehart, and Winston, 1982.

Kortlandt, Adriaan. "Chimpanzees in the Wild." *Scientific American* 206 (May 1962): 128–38.

Krutch, Joseph Wood. *The Best of Two Worlds.* New York: William Sloane Associates, 1950.

———. *The Great Chain of Life.* Boston: Houghton Mifflin, 1956.

Kruuk, Hans. *The Spotted Hyena: A Study of Predation and Social Behavior.* Chicago: University of Chicago Press, 1972.

Kummer, Hans. *Social Organization of Hamadryas Baboons; A Field Study.* Chicago and London: University of Chicago Press, 1968.

Laidler, Keith. *The Talking Ape.* New York: Stein & Day, 1980.

Lavrov, L. S. "Evolutionary Development of the Genus *Castor* and Taxonomy of the Contemporary Beavers of Eurasia." *Acta Zoologica Fennica* 174 (1983): 87–90.

Lawrence, Elizabeth Atwood. *Rodeo: An Anthropologist Looks at the Wild and the Tame.* Knoxville, TX: University of Texas Press, 1982.

Lazell, James D., Jr., and Spitzer, Numi C. "Apparent Play Behavior in an American Alligator." *Copeia* (1977): 188.

Leighton, Donna Robbins. "Gibbons: Territoriality and Monogamy." In *Primate Societies,* edited by Barbara B. Smuts, Dorothy L. Cheney, Robert M. Seyfarth, Richard W. Wrangham, and Thomas T. Struhsaker. Chicago and London: University of Chicago Press, 1986.

Lewis, George, with Fish, Byron. *Elephant Tramp.* Boston: Little, Brown and Co., 1955.

Lewis, Michael. *Shame: The Exposed Self.* New York: The Free Press/Macmillan, 1992.

Leyhausen, Paul. *Cat Behavior: The Predatory and Social Behavior of Domestic and Wild Cats.* Translated by Barbara A. Tonkin. New York and London: Garland STPM Press, 1979.

Linden, Eugene. *Silent Partners: The Legacy of the Ape Language Experiments.* New York: Times Books, 1986.

Lopez, Barry Holstun. *Of Wolves and Men.* New York: Charles Scribner's Sons, 1987.

Lorenz, Konrad. *The Year of the Greylag Goose.* New York and London: Harcourt Brace Jovanovich, 1978.

Lutts, Ralph H. *The Nature Fakers; Wildlife, Science and Sentiment.* Golden, CO: Fulcrum, 1990.

Macdonald, David. *Running with the Fox.* London and Sydney: Unwin Hyman, 1987.

Mader, Troy R. "Wolves and Hunting." *Abundant Wildlife,* Special Wolf Issue (1992): 3.

Magel, Charles R. *Bibliography of Animal Rights and Related Matters.* University Press of America, 1981.

Magoun, A. J., and Valkenburg, P. "Breeding Behavior of Free-ranging Wolverines *(Gulo).*" *Acta Zoologica Fennica* 174 (1983): 149–51.

Mahaffy, J. P. *Descartes.* Edinburgh: Blackwood, 1901.

Mansergh, Ian and Broome, Linda. *The Mountain Pygmy-possum of the Australian Alps.* Kensington, NSW, Australia: New South Wales University Press, 1994.

Martin, Esmond, and Martin, Chrysse Bradley. *Run Rhino Run.* London: Chatto and Windus, 1982.

Masserman, Jules H.; Wechkin, Stanley; and Terris, William. " 'Altruistic' Behavior in Rhesus Monkeys." *American Journal of Psychiatry* 121 (1964): 584–85.

Mayes, Andrew. "The Physiology of Fear and Anxiety." In *Fear in Animals and Man,* edited by W. Sluckin, 24–55. New York and London: Van Nostrand Reinhold Co., 1979.

McFarland, David, ed. *The Oxford Companion to Animal Behavior.* Oxford and New York: Oxford University Press, 1987.

McNulty, Faith. *The Whooping Crane: The Bird That Defies Extinction.* New York: E. P. Dutton & Co., 1966.

"Medicine and the Media." Editorial. British Medical Journal 295 (September 12, 1987), 659–60.

Midgley, Mary. "The Concept of Beastliness: Philosophy, Ethics and Animal Behavior." *Philosophy* 48 (1973): 111–35.

———. *Beast and Man: The Roots of Human Nature.* Ithaca, NY: Cornell University Press, 1978.

———. *Animals and Why They Matter.* Athens, GA: University of Georgia Press, 1983.

———. "The Mixed Community." In *The Animal Rights/Environmental Ethics Debate,* edited by Eugene C. Hargrove. Albany: State University of New York, 1992.

Millay, Edna St. Vincent. *Collected Lyrics.* New York: Washington Square Press, 1959.

Mitchell, Robert W., and Thompson, Nicholas S. *Deception: Perspectives on Human and Nonhuman Deceit.* Albany: State University of New York Press, 1986.

Moggridge, J. Traherne. *Harvesting Ants and Trap-Door Spiders: Notes and Observations on Their Habits and Dwellings.* London: L. Reeve & Co., 1873.

Monastersky, Richard. "Boom in 'Cute' Baby Dinosaur Discoveries." *Science News* 134 (October 22, 1988): 261.

Montaigne, Michel. *The Complete Works of Montaigne.* Translated by D. M. Frame. Vol. 2. Garden City, NY: Anchor Books, 1960.

Montgomery, Sy. *Walking with the Great Apes.* Boston: Houghton Mifflin, 1991.

Moore, J. Howard. *The Universal Kinship.* 1906; reprint, Sussex, England: Centaur Press, 1992.

Morey, Geoffrey. *The Lincoln Kangaroos.* Philadelphia: Chilton Books, 1963.

Morris, Desmond. *The Biology of Art: A Study of the Picture-Making Behavior of the Great Apes and Its Relationship to Human Art.* New York: Alfred A. Knopf, 1962.

———. *Animal Days.* London: Jonathan Cape, 1979; New York: Perigord Press/ William Morrow and Co., 1980.

Morton, Eugene S., and Page, Jake. *Animal Talk: Science and the Voices of Nature.* New York: Random House, 1992.

Moss, Cynthia. *Portraits in the Wild: Behavior Studies of East African Mammals.* Boston: Houghton Mifflin Company, 1975.

———. *Elephant Memories: Thirteen Years in the Life of an Elephant Family.* New York: William Morrow and Co., 1988.

Nathanson, Donald. *Shame and Pride: Affect, Sex, and the Birth of the Self.* New York: W. W. Norton & Company, 1992.

Nishida, Toshisada. "Local Traditions and Cultural Transmission." In *Primate Societies.* See Leighton, 1986.

Nollman, Jim. *Animal Dreaming: The Art and Science of Interspecies Communication.* Toronto and New York: Bantam Books, 1987.

Norris, Kenneth S. *Dolphin Days: The Life and Times of the Spinner Dolphin.* New York and London: W. W. Norton & Co., 1991.

Ogden, Paul. *Chelsea: The Story of a Signal Dog.* Boston: Little, Brown and Co., 1992.

Orleans, R. Barbara. *In the Name of Science: Issues in Responsible Animal Experimentation.* New York: Oxford University Press, 1992.

Packer, Craig. "Male Dominance and Reproductive Activity in *Papio anubis.*" *Animal Behavior* 27 (1979): 37–45.

Patenaude, Françoise. "Care of the Young in a Family of Wild Beavers, *Castor canadensis.*" *Acta Zoologica Fennica* 174 (1983): 121–22.

Patterson, Francine, and Linden, Eugene. *The Education of Koko.* New York: Holt, Rinehart & Winston, 1981.

Patterson, Francine. *Gorilla: Journal of the Gorilla Foundation* 15, #2 (June 1992).

Paulsen, Gary. *Winterdance: The Fine Madness of Running the Iditarod.* New York: Harcourt Brace & Co., 1994.

Plotnicov, Leonard. "Love, Lust, and Found in Nigeria." Paper presented at the 1992 American Anthropological Association annual meeting, San Francisco, December 2, 1992.

Premack, D., and Woodruff, G. "Does the Chimpanzee Have a Theory of Mind?" *Behavior and Brain Science* 1 (1978): 515–26.

Pryor, Karen. *Lads Before the Wind: Adventures in Porpoise Training.* New York: Harper & Row, 1975.

Pryor, Karen, and Norris, Kenneth S. *Dolphin Societies: Discoveries and Puzzles.* Berkeley, CA: University of California Press, 1991.

Pryor, Karen; Haag, Richard; and O'Reilly, Joseph. "The Creative Porpoise: Training for Novel Behavior." *Journal of the Experimental Analysis of Behavior,* 12 (1969): 653–61.

Rachels, J. *Created from Animals: The Moral Implications of Darwinism.* Oxford: Oxford University Press, 1990.

Rajecki, D. W., ed. *Comparing Behavior: Studying Man Studying Animals.* Hillsdale, NJ: Lawrence Erlbaum Associates, 1983.

Rasa, Anne. *Mongoose Watch: A Family Observed.* Garden City, NY: Anchor Press/ Doubleday & Co., 1986.

Reed, Don C. *Notes from an Underwater Zoo.* New York: Dial Press, 1981.

Regan, Tom. *The Case for Animal Rights.* Berkeley, CA: University of California Press, 1983.

Regan, Tom, and Singer, Peter, eds. *Animal Rights and Human Obligations.* Englewood Cliffs, NJ: Prentice-Hall, 1976.

Reinhold, Robert. "At Sea World, Stress Tests Whale and Man." *New York Times,* April 4, 1988: A9.

Ristau, Carolyn A., ed. *Cognitive Ethology: The Minds of Other Animals (Essays in Honor of Donald R. Griffin).* New Jersey: Lawrence Erlbaum Associates, 1991.

Roberts, Catherine. *The Scientific Conscience: Reflections on the Modern Biologist and Humanism.* New York: George Braziller, 1967.

Romanes, George J. *Animal Intelligence.* London: Kegan Paul, Trench, Trubner and Co., 1882.

———. *Mental Evolution in Animals.* London: Kegan Paul, Trench, Trubner and Co., 1883.

Rosenfield, Leonora Cohen. *From Beast-Machine to Man-Machine: Animal Soul in French Letters from Descartes to La Mettrie.* 1940; new edition, New York: Octagon Books, 1968.

Rowell, Thelma. *The Social Behaviour of Monkeys.* Harmondsworth, Middlesex, England: Penguin Books, 1972.

Rowley, Ian, and Chapman, Graeme. "Cross-fostering, Imprinting and Learning in Two Sympatric Species of Cockatoo." *Behaviour* 96 (1986): 1–16.

Rozin, Paul, and Fallon, April. "A Perspective on Disgust." *Psychological Review* 94 (1987): 23–41.

Rupke, Nicolaas A., ed. *Vivisection in Historical Perspective.* London: Croom Helm, 1987.

Russell, Diana E. H. *The Politics of Rape: The Victim's Perspective.* New York: Stein & Day, 1977.

———. *Rape in Marriage.* New York: Macmillan, 1982.

———. "The Incidence and Prevalence of Intrafamilial and Extrafamilial Sexual Abuse of Female Children." *Child Abuse and Neglect: The International Journal* 7 (1983): 133–46.

———. *The Secret Trauma: Incestuous Abuse of Women and Girls.* New York: Basic Books, 1986.

Russell, Diana E. H., and Howell, Nancy. "The Prevalence of Rape in the United States Revisited." *Signs: Journal of Women in Culture and Society* 8 (Summer 1983): 668–95.

Russell, P. A. "Fear-Evoking Stimuli." In *Fear in Animals and Man.* See Mayes, 1979.

Rutter, Russell J. and Pimlott, Douglas H. *The World of the Wolf.* Philadelphia and New York: J. B. Lippincott Co., 1968.

Ryden, Hope. *God's Dog.* New York: Coward, McCann & Geoghegan, 1975.

———. *Lily Pond: Four Years with a Family of Beavers.* New York: William Morrow & Co., 1989.

Sadoff, Robert L. "The Nature of Crying and Weeping." In *The World of Emotion: Clinical Studies of Affects and Their Expression,* edited by Charles W. Socarides. New York: International Universities Press, 1977.

Savage, E. S.; Temerlin, Jane; and Lemmon, W. B. "The Appearance of Mothering Behavior Toward a Kitten by a Human-Reared Chimpanzee." Paper delivered at the Fifth Congress of Primatology, Nagoya, Japan, 1974.

Savage-Rumbaugh, E. Sue. *Ape Language: From Conditioned Response to Symbol.* New York: Columbia University Press, 1986.

Schaller, George B. *The Serengeti Lion: A Study of Predator-Prey Relations.* Chicago and London: University of Chicago Press, 1972.

———. *The Last Panda.* Chicago and London: University of Chicago Press, 1993.

Schechter, Neil; Berde, Charles B.; and Yaster, Myron, eds. *Pain in Infants, Children, and Adolescents.* Baltimore: Williams and Wilkins, 1993.

Scheffer, Victor B. *Seals, Sea Lions, and Walruses: A Review of the Pinnipedia.* Stanford, CA: Stanford University Press, 1958.

Schiller, Paul H. "Figural Preferences in the Drawings of a Chimpanzee." *Journal of Comparative and Physiological Psychology* 44 (1951): 101–11.

Schullery, Paul. *The Bear Hunter's Century.* (New York: Dodd, Mead & Co., 1988).

Seligman, Martin E. P. *Helplessness: On Depression, Development, and Death.* San Francisco: W. H. Freeman & Co., 1975.

Seyfarth, Robert M., and Cheney, Dorothy L. "Grooming, Alliances, and Reciprocal Altruism in Vervet Monkeys." *Nature* 308, #5 (April 1984): 541–42.

Sidowski, J. B. "Psychopathological Consequences of Induced Social Helplessness During Infancy." In *Experimental Psychopathology: Recent Research and Theory*, edited by H. D. Kimmel. New York: Academic Press, 1971.

Singer, Peter. *Animal Liberation*. New York Review, 1975.

Singh, Arjan. *Tiger! Tiger!* London: Jonathan Cape, 1984.

Small, Meredith F., ed. *Female Primates: Studies by Women Primatologists*. New York: Alan R. Liss, 1984.

Smith, J. Maynard, and Ridpath, M. G. "Wife Sharing in the Tasmanian Native Hen, *Tribonyx mortierii:* A Case of Kin Selection?" *The American Naturalist* 106 (July–August 1972): 447–52.

Smuts, Barbara. "Dominance: An Alternative View." *Behavioral and Brain Sciences* 4 (1981): 448–49.

Spiegel, Marjorie. *The Dreaded Comparison: Human and Animal Slavery*. Philadelphia: New Society Publishers, 1988.

Staddon, J. E. R. "Animal Psychology: The Tyranny of Anthropocentrism." In *Whither Ethology? Perspectives in Ethology*, P. P. G. Bateson and Peter H. Klopfer, eds. New York: Plenum Press, 1989.

Starobinski, Jean. "Rousseau et Buffon." *Gesnerus* 21 (1964): 83–94.

Strum, Shirley C. *Almost Human: A Journey into the World of Baboons*. New York: Random House, 1987.

Symons, Donald. *The Evolution of Human Sexuality*. New York: Oxford University Press, 1979.

Teal, John J., Jr. "Domesticating the Wild and Woolly Musk Ox." *National Geographic* (June 1970).

Terrace, Herbert. *Nim: A Chimpanzee Who Learned Sign Language*. New York: Washington Square Press, 1979.

Thomas, Elizabeth Marshall. "Reflections: The Old Way." *The New Yorker* (October 15, 1990): 78–110.

———. *The Hidden Life of Dogs*. Boston and New York: Houghton Mifflin Co., 1993.

———. *The Tribe of Tiger*. New York: Simon & Schuster, 1994.

Thomson, Robert. "The Concept of Fear." In *Fear in Animals and Man*. See Mayes, 1979.

Tomkies, Mike. *On Wing and Wild Water*. London: Jonathan Cape, 1987.

———. *Last Wild Years*. London: Jonathan Cape, 1992.

Trivers, Robert L. "The Evolution of Reciprocal Altruism." *Quarterly Review of Biology* 46 (1971): 35–57.

Turner, E. S. *All Heaven in a Rage*. Sussex, England: Centaur Press, 1992.

Turner, J. *Reckoning with the Beast: Animals, Pain, and Humanity in the Victorian Mind*. Baltimore: The Johns Hopkins University Press, 1980.

Tyack, Peter. "Whistle Repertoires of Two Bottle-nosed Dolphins, *Tursiops truncatus:* Mimicry of Signature Whistles?" *Behavioral Ecology and Sociobiology* 18 (1989): 251–57.

Voltaire, François-Marie Arouet de. *Dictionnaire philosophique*, edited by Julien Benda and Raymond Naves. Paris: Garnier Frères, 1961.

———. "The Beasts." Article 6 in *Le philosophe ignorant*. Les Oeuvres Complètes de Voltaire, Vol. Mélanges, edited by Jacques van den Heuvel. Paris: Gallimard.

Walker, Ernest P. *Mammals of the World,* 2nd ed. Baltimore: The Johns Hopkins Press, 1986.

Walker, S. *Animal Thought.* London: Routledge & Kegan Paul, 1983.

Welty, Joel Carl, and Baptista, Luis. *The Life of Birds,* 4th ed. New York: Saunders College Publishing, 1988.

Wierzbicka, Anna. "Human Emotions: Universal or Culture-Specific?" *American Anthropologist* 88 (1986): 584–94.

Wiesner, Bertold P., and Sheard, Norah M. *Maternal Behavior in the Rat.* Edinburgh and London: Oliver & Boyd, 1933.

Wigglesworth, V. B. "Do Insects Feel Pain?" *Antenna* 4 (1980): 8–9.

Wilkinson, Gerald S. "Food Sharing in Vampire Bats." *Scientific American* 262 (1990): 76–82.

Williams, J. H. *Elephant Bill.* Garden City, NY: Doubleday & Co., 1950.

Wilsson, Lars. *My Beaver Colony.* Translated by Joan Bulman. Garden City, NY: Doubleday & Co., 1968.

Wiltschko, Wolfgang; Munro, Ursula; Ford, Hugh Ford; and Wiltschko, Roswitha. "Red Light Disrupts Magnetic Orientation of Migratory Birds." *Nature* 364 (August 5, 1993): 525.

Winslow, James T.; Hastings, Nick; Carter, C. Sue; Harbaugh, Carroll R.; and Insel, Thomas R. "A Role for Central Vasopressin in Pair Bonding in Monogamous Prairie Voles." *Nature* 365 (October 7, 1993): 545–48.

Wittgenstein, Ludwig. *Philosophical Investigations,* 3rd ed. Translated by G. E. M. Anscombe. New York: Macmillan Co., 1968.

Wu, Hannah M. H.; Holmes, Warren G.; Medina, Steven R.; and Sackett, Gene P. "Kin Preference in Infant *Macaca nemestrina.*" *Nature* 285 (1980): 225–27.

Yerkes, Robert M., and Yerkes, Ada W. *The Great Apes: A Study of Anthropoid Life.* New Haven, CT: Yale University Press, 1929.

Young, Stanley P. *The Wolves of North America: Their History, Life Habits, Economic Status, and Control.* Part II: "Classification of Wolves," by Edward A. Goldman. Washington, DC: American Wildlife Institute, 1944.

Index